RESEARCH METHODS FOR SPORTS PERFORMANCE ANALYSIS

Modern techniques of sports performance analysis enable the sport scientist, coach and athlete to objectively assess, and therefore improve upon, sporting performance. They are an important tool for any serious practitioner in sport and, as a result, performance analysis has become a key component of degree programmes in sport science and sports coaching.

Research Methods for Sports Performance Analysis explains how to undertake a research project in performance analysis including:

- selection and specification of a research topic
- the research proposal
- gaining ethical approval for a study
- developing a performance analysis system
- testing a system for reliability
- analysing and discussing data
- writing up results.

Covering the full research cycle and clearly introducing the key themes and issues in contemporary performance analysis, this is the only book that sports students will need to support a research project in performance analysis, from undergraduate dissertation to doctoral thesis. Including case studies, examples and data throughout, this book is essential reading for any student or practitioner with an interest in performance analysis, sports coaching or applied sport science.

Peter O'Donoghue is Reader and Discipline Director for Performance Analysis in the Cardiff School of Sport, UWIC. He is also the editor of the *International Journal of Performance Analysis of Sport*.

RESEARCH METHODS FOR SPORTS PERFORMANCE ANALYSIS

Peter O'Donoghue

Routledge
Taylor & Francis Group

LONDON AND NEW YORK

Published 2010
by Routledge
2 Park Square, Milton Park, Abingdon, Oxon, OX14 4RN

Simultaneously published in the USA and Canada
by Routledge
270 Madison Avenue, New York, NY 10016

Routledge is an imprint of the Taylor & Francis Group, an informa business

© 2010 Peter O'Donoghue

Typeset in Sabon and Futura by
Saxon Graphics
Printed and bound in Great Britain by
CPI Antony Rowe, Chippenham, Wiltshire

British Library Cataloguing in Publication Data
A catalogue record for this book is available from the British Library

Library of Congress Cataloging-in-Publication Data
O'Donoghue, Peter.
Research methods for sports performance analysis / Peter O'Donoghue.
p. cm.
Includes bibliographical references.
1. Sports sciences—Research. 2. Sports—Research. 3. Sports—Study and teaching.
4. Sports—Physiological aspects. I. Title.
GV558.0375 2010
612'.044—dc22   2009036584

ISBN10: 0-415-49622-5 (hbk)
ISBN10: 0-415-49623-3 (pbk)
ISBN10: 0-203-87830-2 (ebk)

ISBN13: 978-0-415-49622-3 (hbk)
ISBN13: 978-0-415-49623-0 (pbk)
ISBN13: 978-0-203-87830-9 (ebk)

CONTENTS

FIGURES

TABLES

ACKNOWLEDGEMENTS

I'd like to express my sincere thanks to all of the people who helped me during the writing and production of this book. In particular, I'd like to thank Brian Guerin and Simon Whitmore from Routledge and my colleagues in the Centre for Performance Analysis at the University of Wales Institute Cardiff.

PREFACE

There is a series of important textbooks that have been written about performance analysis of sport and published by Routledge. The primary texts are *The Essentials of Performance Analysis of Sport* (Hughes and Franks, 2008), the second edition of *Notational Analysis of Sport* (Hughes and Franks, 2004) and *Handbook of Soccer Match Analysis* (Carling *et al.*, 2005). The textbooks edited by Hughes and Franks are excellent secondary sources covering system development, research topics and the underlying area of feedback and communication in coaching contexts. Carling *et al.*'s book was targeted at a practitioner audience and covers the use of up-to-date systems in professional soccer analysis. There were many reasons for the author wishing to write this current textbook. First, three main textbooks is a very limited base of secondary literature for the growing number of performance analysis undergraduate and postgraduate students to use. The leading academics in performance analysis are approaching retirement but the academic work in the subject must continue, especially with the rapid growth in the performance analysis profession over the last five years.

Secondly, the author has spent the last 10 years at the University of Ulster and the University of Wales Institute Cardiff giving similar advice to performance analysis project students that cannot be found in the literature. There is a clear gap in the literature for a textbook that the students can use during a research project in performance analysis. This book covers all of the stages of a performance analysis research project, from selecting a research topic of interest to delivery of the completed dissertation.

The third reason for producing this textbook at this time is that there are fundamental changes being proposed by the author to the field of performance analysis. These changes relate to the underlying ontological and epistemological assumptions of research methods used in performance analysis. The traditional view of performance analysis is that it involves objective methods that are based on the assumptions of the normative paradigm. Where human operators form part of any observational analysis system, there will always be an element of subjective classification of observed behaviour. Many students in the past have been given an impossible task to produce precise, unambiguous operational definitions of events and behaviours. The author is, therefore, proposing that these subjective processes are recognised within performance analysis methodology. Performance analysis has been characterised in the past as including biomechanics and notational analysis. It is the view of this author that any research that involves the analysis of actual sports performance falls under the umbrella of performance analysis of sport. This includes qualitative analysis of sports performance and the use of heart rate and other physiological data recorded during actual sports performance. Therefore, both quantitative and qualitative methods can be applied in performance analysis and these are based on the assumptions of the normative and interpretive paradigms respectively.

In recent years, the author has undertaken research work in performance prediction that involves forming models from past data to predict the outcomes of future matches. The author applied these techniques when working with a netball squad to show that the squad was expected to lose an important match. However, the whole point of analysing performance is to identify areas requiring attention and to support decisions about match preparation that will result in a more successful outcome. Thus the squad are in a continuous process of change and past performances do not dictate the outcomes of future matches. The role of the performance analyst working with coaches and athletes is to help enhance performance through a cycle of observation, analysis, reflection, planning and action. The assumptions of critical theory and action research are relevant to this process, which challenges the status quo that a team is expected to lose a forthcoming match. There is still some scepticism about the value of performance analysis support and, therefore, intervention studies into the effectiveness of instructional and motivational feedback are needed. These studies should be done with an understanding of the underlying assumptions of the action research methods used.

There are some terms used throughout the text that come from the author's current role as a university lecturer in the UK higher education system. Levels 2, 3 and 4 refer to the second year of an undergraduate (Bachelor's) degree programme, the final year of an undergraduate programme and Master's level work respectively. The description of the research process experienced by students is also influenced by the processes used at the author's current university. Research proposals are used within a level 2 research methods module primarily to assess learning outcomes for research methods but also to help students make a start on their level 3 research projects. This allows for allocation of supervisors and ethical consideration of the research proposals prior to the student's final year commencing. Many universities adopt a similar approach although some may separate dissertation proposals from the teaching of research methods. The terms 'sports studies', 'sports science', and 'sport and exercise science' are often used interchangeably to describe the wider area within which performance analysis exists. The text often refers to the student but sometimes refers to the researcher, particularly when more advanced material is being covered, for example applying for external research funding.

The author often uses his own MSc Sport, Exercise and Leisure dissertation as an example of different parts of the research process. This is because this particular research project was an experience of what project students in sports science go through. The project is often referred to through the paper that was published in the *Journal of Sports Sciences* (O'Donoghue and Ingram, 2001), which may seem strange to readers as the passages often describe what the author did when referring to this co-authored paper. My co-author was my dissertation supervisor, Billy Ingram, who played a vital role, generously offering advice that was gratefully received. It is hoped that this experience of personally undertaking a research project as part of a university Master's programme will convince the reader that this book has been written by an author acquainted with research projects not only as a supervisor but also as a student.

CHAPTER 1

PERFORMANCE ANALYSIS RESEARCH

INTRODUCTION

This first chapter introduces performance analysis research in four parts:

1. An introduction to performance analysis of sport.
2. An introduction to performance analysis research.
3. History of performance analysis research.
4. An introduction to research topics in performance analysis.

In introducing the area of performance analysis of sport, there are six questions that should be answered:

> What is performance analysis of sport?
> Why do we do performance analysis of sport?
> Who does performance analysis of sport?
> Where is performance analysis of sport done?
> When is performance analysis of sport done?
> How is performance analysis of sport done?

Performance analysis is an area of sport and exercise science concerned with actual sports performance rather than self-reports by athletes or laboratory experiments. The applied nature of performance analysis research is described and a justification for performance analysis as a research area is made. There is overlap between performance analysis and other disciplines as technical, physical and psychological aspects of performance are often being investigated within performance analysis investigations. The scope of this book does not cover these other areas, but it is recognised that this overlap exists and that often this book would be used in conjunction with other research methods material by students undertaking such projects. This introductory chapter provides a brief history of performance analysis of sport as well as other relevant developments that have had an impact on the area. There are many types of study that can be done and that students are encouraged to undertake, too many to fully list in this textbook. However, the hot topics in performance analysis research at the time of this first edition being written are covered at the end of this opening chapter.

PERFORMANCE ANALYSIS OF SPORT: WHAT? WHY? WHO? WHERE? WHEN? HOW?

What is performance analysis of sport?

Performance analysis of sport is the investigation of actual sports performance or performance in training. What distinguishes performance analysis from other disciplines is that it is concerned with actual sports performance rather than activity undertaken in laboratory settings or data gathered from self-reports such as questionnaires, focus groups, accounts and interviews. There are cases where laboratory-based biomechanics exercises can count as performance analysis; if the technique under investigation is an important skill within the sport of interest, then detailed biomechanical analysis of the technique is analysis of that skill. This is an especially strong argument where a skill such as running stride, golf swing or tennis serve is critically important to success in the sport and where the detailed data required cannot be gathered during actual competition.

When Hughes and Bartlett (2008) wrote a book chapter entitled 'What is performance analysis?' they covered notational analysis and biomechanics. The view of this author is that all research that involves the analysis of actual sports performance in training or competition can be referred to as 'performance analysis'. Physiological variables such as heart rate response and blood lactate accumulation can be gathered in many situations including sports competition where the nature of the sport allows such measurement to be made. The fact that the measurements relate to sports performance, rather than a fitness test or laboratory-based tests or exercises, ensures that such investigations fall under the umbrella of performance analysis of sport. The use of questionnaire instruments and other self-reports within performance analysis is sometimes possible where such instruments have been validated against other gold standard measurements. The profile of mood states inventory (McNair *et al.*, 1971) or the rate of perceived exertion scale (Borg, 1982) are examples of self-reports that can be used periodically during sports performance to record useful information about the performance.

Why do we do performance analysis of sport?

The main reason for doing performance analysis is to develop an under-standing of sports that can inform decision making by those seeking to enhance sports performance. The complexities and dynamic nature of many sports means that observation and measurement is needed to improve our understanding of performance. Of course, coaches and physical education teachers have observed and analysed performance subjectively as long as sport has existed. However, such observations have their limitations. Franks and Miller (1991) found that international level soccer coaches recalled 45 per cent of these key factors correctly. More recently, a study took place of the recall ability of eight adult qualified football coaches with a minimum qualification of level one in one or more of the four Scottish Football Association accredited coaching categories at least six months experience after obtaining their qualification (Laird and Waters, 2008). These coaches accurately recalled 59 per cent of the critical events within a 45-minute period of soccer play. Franks (1993) did a study of experienced and inexpe-rienced gymnastics coaches' perceptions of two performances. Experienced coaches were more likely to report there was a difference when there was no difference between the performances and did not identify actual differences in the performances any more successfully than the inexperienced coaches. Maslovat and Franks (2008) summarised some studies of eyewitness state-ments of crime where some of the explanations of inaccuracies may be rel-evant to subjective observation of sports performance. These explanations include increased arousal level, observer bias and errors in attentional focus. The rationale for using performance analysis is to overcome the limitations of using subjective observation alone and to provide objective information to achieve a greater understanding of the performance. This information, in turn, assists decision making by coaches and may, therefore, play a vital role in performance enhancement. Maslovat and Franks (2008) described the coaching process according to Franks *et al.* (1983) and how it is enhanced through performance analysis support. There are other non-coaching uses of performance analysis of sport, for example in the media and the judging of performances where performance analysis also assists decision making.

The reasons for using performance analysis can be explored in more depth when examining the purposes of performance analysis. Different authors have identified different sets of purposes of notational analysis and biomechanics. Hughes (1986) identified the four purposes of notational analysis as technical evaluation, tactical evaluation, analysis of movement and statistical compilation. Statistical compilation is something that over-laps with the other three purposes and so the purposes of notational analy-sis were revised to the following five purposes of notational analysis (Hughes, 1998): technical evaluation, tactical evaluation, analysis of movement, coach and player education, and performance modelling using match analy-sis databases.

Hughes and Bartlett (2008) described the similarities between notational analysis and biomechanics including their common purpose of drawing lessons from performances to improve future performances. Biomechanics allows the detail of good and poor technique to be studied. Rather than merely identifying what techniques need to improve, biomechanics helps determine the ways in which techniques need to improve. Biomechanics also has a role in identifying technique with potential for injury (Bartlett, 1999: 1–145, Elliott *et al.*, 1996). Lees (2008) described how technique analysis involves a fault identification and diagnosis stage as well as a remediation stage. While there are differences between biomechanics and notational analysis, one of their similarities is the purpose of helping improve performance. The two stages of technique analysis described by Lees (2008) are very similar to the way in which notational analysis and notational analysis-based feedback are integrated within coaching processes (Franks *et al.*, 1983).

Other purposes of performance analysis can be recognised through research work that has taken place, the areas of interest listed by academic conferences, and topics written about in performance analysis textbooks. Carling *et al.* (2005: 11–12) listed tactical, technical, physical movement and behavioural aspects as well as critical incidents as areas to analyse within match analysis of soccer. The most recent two World Congresses of Performance Analysis of Sport were the 7th (Szombathely, Hungary, 2006) and 8th (Magdeburg, Germany, 2008) in the series. The topics covered in these conferences were:

- neuromotor control;
- neural networks;
- motor learning and feedback;
- exercise performance testing/analysis of elite athletes/referees;
- coaching process;
- coach behaviour;
- computational analysis;
- sports performance database management;
- sports technology;
- performance analysis in health and senior sports;
- analysis of movement;
- notational analysis;
- sports software systems;
- working with elite performers;
- sports science and performance.

Some of these topics are to do with methods of analysis rather than actual performance analysis purposes. For example neural networks, computational methods, sports software systems and sports technology would not be purposes of performance analysis although presentations on methodo-

Table 1.1 Purposes of performance analysis relevant to the analysis of different people

Area being analysed	Players	Coaches	Referees
Technique	YES		
Technical effectiveness	YES		
Tactical evaluation, decision making	YES	YES	YES
Movement	YES		YES
Behaviour/psychological aspects	YES	YES	YES

logical and technological developments are key contributions within the World Congresses. Within this text, a slightly revised set of purposes of performance analysis is proposed indicating how the purposes apply to the analysis of players, coaches and referees, as shown in Table 1.1.

Technique analysis includes exercise performance testing where key skills of the sport are being investigated. These purposes can apply to the investigation of different age groups, elite participants, different levels of participants or recreational participants. Therefore, the topics such as elite performers and other populations would not be included in the purposes of performance analysis listed here. Injury risk is included within analysis of movement and analysis of technique. The coaching process topic would not qualify as a purpose of performance analysis in its own right as this topic is investigated through various analyses of coaches and players. The purposes of notational analysis of statistical compilation (Hughes, 1986), coach and player education (Hughes, 1998), the development of performance databases and performance modelling (Hughes, 1998) are very broad purposes of performance analysis that include different specific purposes. For example when performance databases are developed, those databases will include technique, technical effectiveness, tactical, behavioural, movement and decision making data. Motor control and neuromotor control are not purposes of performance analysis but explanations of technique.

Technical effectiveness does not look at the technique itself to determine flaws in how a skill was performed. Instead, skills are assessed through positive to negative ratios allowing those skills requiring attention to be identified. An example of this are the frequency profiles of defensive and offensive positive and negative applications of different skills in soccer devised by the Norwegian Football Association (Olsen and Larsen, 1997). An example from racket sports would be frequency profiles of winners and errors (Murray and Hughes, 2001).

Technique analysis considers the mechanical aspects of technique and is concerned with the way in which the skill is performed in terms of kinetic and kinematic detail of the movement involved. This has been done without the aid of expensive biomechanical analysis equipment using qualitative analysis of mechanical properties of sports movement while analysing video recordings of performance (Underwood and McHeath, 1977). Chapter 16 of McGinnis's (1999) biomechanics textbook describes a process of qualitative biomechanics to improve the performance of a skill. The coach uses his or her knowledge of the theoretical model of the most effective technique to achieve the goal of the skill. The coach observes the athlete performing the skill from an effective viewing position. This is done during training and it is important to duplicate competition conditions during the exercise. The observed performance is evaluated by comparing it with the theoretical ideal performance. Errors in the performance of the skill are identified considering the morphological constraints of the performer as well as the influence of any equipment and environmental problems. The errors are considered in terms of any injury risk, their effect on achieving the goal of the skill, where they occur during the performance of the skill and how difficult the errors are to correct. Once the performance has been observed and evaluated, the coach provides the instruction to the athlete that is necessary to correct the errors in technique. The feedback includes what the athlete did incorrectly and can be provided verbally with the aid of demonstrations. Clear instruction is then given on what the coach wants the athlete to do and drills are devised to help the athlete to correct any errors. This process of feedback and instruction is repeated throughout training sessions where the set drills are performed.

Tactical analysis is concerned with strategy and tactics. A strategy can be thought of as a plan that has been established prior to competition that will make best use of the player's or team's strengths while limiting the effects of any weaknesses. At the same time, the strategy should seek to exploit any known weaknesses of the opposition while avoiding situations where the opposition can make use of their strengths. Tactics on the other hand are moment-to-moment decisions made during the competition by players based on the options available to them and the perceived risks and opportunities associated with each (Fuller and Alderson, 1990). The strategy that has been decided before the match and tactical decisions made during the match are not directly observable during competition. However, the different skills performed by players, the locations where they are performed as well as the timing of these actions can give an indication of the strategy and tactics being applied. For example, if a tennis player approached the net then the player can be assumed to have adopted a net strategy. If, on the other hand, the player does not approach the net, then the player can be assumed to have adopted a baseline strategy. An extension of tactical analysis is analysis of decision making in general through observational means. Decisions by players and officials can be investigated indirectly through observational means. The analysis of referees' decisions is easier than the analysis of player decisions. This is because there are regulations of play and the performance information will allow analysis of whether those regulations were applied correctly by the official. The analysis of player decisions requires an understanding of different options available, their relative chance of success and any risks involved. The time pressure that a player was under when making a decision is something that may also be considered when analysing the performance of players. Some decisions of coaches can be done during observation of performance; for example substitutions and the use of time-outs.

The analysis of movement is important to gain an understanding of the physical demands of sports, tactical elements as well as injury risk within those sports (O'Donoghue, 2008b). Time-motion analysis is concerned with player movement throughout the entire game and not just on the ball movement. This has allowed an estimation of the energy systems involved in racket sports (Richers, 1995) and field games (Spencer et al., 2004). The data used in time-motion analysis can be the distribution of time among different activities, distances covered and velocity profiles as well as supplementary heart rate and blood lactate data (Bangsbo, 1993: 36–40, Bangsbo, 1997).

The idea that tactics can be inferred from observed patterns of behaviour can also be extended to psychological aspects of performance such as alertness, arousal, aggression and attentional focus. In mainstream psychology, observational techniques are used in areas such as aggression (Dadds et al., 1996) and integrated video analysis systems such as Observer Pro can also be applied to various analyses of human behaviour (Noldus, 2009). Observational studies such as that of Dadds et al. (1996) are not restricted to the analysis of action, but also use conversations that are observed and other events associated with communication. In sports performance, one study of attentional focus required a combination of both observational and self-report techniques (O'Donoghue and Scully, 1999). Errors of attentional focus were identified as internal focus being too broad or narrow and external focus being too narrow or too broad. External focus is based on cues observable to the player while playing the game. A player may focus narrowly on a particular opponent or the ball when other opposing players also pose a threat. Alternatively, the external focus may be too broad where a player is focusing on players who pose no immediate threat at the expense of focusing on those players who are posing an immediate threat. Internal errors of attentional focus include pondering on worries that are not relevant to the game or over-reflecting on a previous incident within the game that the player no longer has any control over. These internal thoughts cannot be inferred from observing player behaviour which is why O'Donoghue and Scully (1999) supported the use of observational techniques with post-match interviews. Aggression in soccer and its association with performance has been investigated using observational analysis of fouls made and cards shown by the referee (Shafizadeh, 2008). There is a role for observational research within the study of aggression and other psychological aspects of performance, but to date this is an underdeveloped area of performance analysis of sport.

Two over-riding purposes of performance analysis that incorporate the more specific purposes described above are coach and player education and the development of performance databases and performance modelling. Coach and player education is used predominantly within practical applied performance analysis support. However, there is still important theoretical research to be undertaken within this area relating to the effectiveness of such support. Rather than assuming that supporting the coach with feedback based on match analysis will enhance performance, we need evidence to support this theory. Early work to investigate the effectiveness of different types of feedback in squash (Brown and Hughes, 1995, Murray *et al.*, 1998) has provided a valuable insight into the difficulties of conducting such research due to the different training patterns of players and other extraneous variables. More recent research has analysed the effectiveness of performance analysis support in Gaelic football (Martin *et al.*, 2004) and in netball (Gasston, 2004, Jenkins *et al.*, 2007). All of these studies have highlighted the difficulties involved in undertaking such research due to the fact that the studies cannot be controlled like laboratory experiments. However, there is still some scepticism about the effectiveness of performance analysis of sport and so further research of this kind is needed to create a pool of evidence for and against the claim that performance analysis has a role in enhancing sports performance.

Performance modelling allows performances to be explained in technical detail (Herzog, 2007, Nigg, 2007b, c, d, Van den Bogert and Nigg, 2007, Yeadon and King, 2008). Hierarchical models allow the relative contribution of different aspects of movement to be related to outcome indicators in closed skills such as the standing long jump (Hay and Reid, 1988: 251–5) and to scores awarded by judges for somersaulting (Takei, 2007). These models can then be manipulated to determine the most productive ways of improving performance. The ability to successfully produce a predictive model of any sport requires a great deal of knowledge and understanding of the sport. Probabilistic models have been produced to represent serving strategy (Gale, 1971) as well as analyse scoring systems within racket sports (Croucher, 1982). These models use the probability of the server winning a point. However, the probability of the server winning a point is an outcome measure and further work is needed to produce models of winning points in terms of process indicators. O'Donoghue (2002) attempted to use correlation techniques to show the relationships between outcome indicators in tennis and process indicators such as service speed, percentage of points where players approached the net, shot types at the net and shot types at the baseline. These models were inconclusive, showing low correlations between process indicators and outcome indicators. This is possibly because there are many different styles of play that can be used successfully or unsuccessfully. However, modelling efforts to explore the factors most associated with successful performance are an important area of future research.

Who does performance analysis of sport?

Performance analysis is primarily done to provide support for individual athletes as well as squads. Within this coaching context, the objective information is often produced by a professional performance analyst who liaises with the coach as part of a coaching process that involves providing feedback to the players. The performance indicators used to assess and monitor athlete performance can also be used by high performance directors within the planning, management and control of elite performance programmes (Greene, 2008). Performance analysis is also undertaken within the judging of some sports. In order to award marks to figure skaters, gymnasts and divers, it is necessary for judges to observe the performances of the athletes and award marks according to established criteria. The computerised scoring system used in amateur boxing requires a set of five judges to observe the contest, awarding marks to a boxer who throws a scoring punch according to the guidelines that the punch is made with the knuckle part of the glove, striking the front of the head or torso with the full weight of the shoulder behind the punch. Whatever the merits of the computerised scoring system for amateur boxing, or those of the more subjective system that preceded it, the fact that the judges have to observe and analyse the performance as part of the judging task brings this type of activity within the definition of performance analysis of sport.

The television, print and internet media use performance analysis within the reporting of sports contests (James, 2008). Of most relevance to the current book, performance analysis is undertaken as part of academic investigations into sports performance in order to provide knowledge about the demands of the sports, factors associated with success or to explain the behaviour of athletes within sports contests. The academic area of performance analysis of sport is still an emerging area, but many investigations have now gathered and analysed sufficient data to be able to propose theories about sports performance. Dynamic systems theory is one such example of theory that has been used to explain the nature of movement in games (Balague *et al.*, 2005, Lames *et al.*, 2009, McGarry, 2006, McGarry and Franks, 1994, McGarry *et al.*, 1999, McGarry *et al.*, 2002, Palut and Zanone, 2005).

Where is performance analysis of sport done?

With live television broadcasting of elite sport, it is no longer necessary for performance analysis to be undertaken at the game venue, although much performance analysis work is still done at the venue for journalism, coaching and judging purposes. Where performance analysis activity is undertaken away from the competition venue, it can be done within purpose built performance analysis laboratories or more flexibly at team hotels, on the team coach when travelling home from an away match, or the analysts' homes as systems are now available on standard laptop computers. Manufacturers of commercial sports analysis software systems have created versions of their products that can operate on palm-top computers. For example, Elite Sports Analysis has developed a system allowing match events to be entered on a palm-top computer at the match venue, where filming may not be permitted, and integrated with a publicly broadcast video of the match away from the competition venue after the match.

When is performance analysis of sport done?

There are routine stages of performance analysis activity that occur once an established system has been developed and accepted by those using it. The first stage is data gathering, which can be done during or after a performance. The second stage is the analysis of the data, which can be done during competition where efficient systems can produce the result required within or immediately after competition. The third stage of performance analysis is the communication of information to the relevant audience depending on the purpose and context. The communication of results can be done in many different forms; for example the display of relevant video sequences to an athlete or squad of players or as the results section of an academic study. Franks (1997) described feed-forward where a performer can be analysed in training situations and results can inform preparation before the next competition. The performances under investigation are done in training and are still analysed during or after those particular performances. In media contexts the results can be presented during a match interval or after the match or even during a match period if required. When performances are analysed for judging purposes, timely communication of the scores is often required. The requirements for the information depend on when and how it is to be used. This in turn dictates how much time is available for data to be analysed. In some academic contexts, the results are not communicated for years after the performance has taken place while in judging and media contexts, performance information is often needed before the contest has ended.

How is performance analysis of sport done?

There are a variety of methods that can be used to gather data for performance analysis exercises, ranging from highly quantitative biomechanical analysis to qualitative analysis. Notational analysis is a method of recording and analysing dynamic and complex situations such as field games. It allows the data to be gathered in an efficient manner, providing an abstract view of the sport that focuses on the most important information. In early manual notational systems, shorthand symbols and tallies allowed data to be recorded efficiently. More recently, computerised notational analysis systems have followed advances in data entry technology and allowed flexible and highly efficient processing of match data. Automatic data gathering systems allow the ball to be tracked in sports like tennis and cricket. An example of such a system is Hawkeye (described by James, 2008). Carling *et al.* (2008) described contemporary systems used for the tracking of multiple players during soccer matches. These include GPS based systems, automatic video tracking and triangular surveying techniques. Some systems are fully automated while others require some operator activity during data gathering. Qualitative analysis has been used in observational studies in socio-cultural areas of sport and has potential for observational analysis of sports performance. The strength of qualitative analysis is that the data recorded are not restricted to a predefined set of events. When describing what performance analysis is earlier in this chapter, a variety of self-report, laboratory-based and physiological measurements were also included within the broad methods that could be used, provided they are used to analyse actual sports performance or, in laboratory settings, techniques that are performed within the sport of interest.

PERFORMANCE ANALYSIS RESEARCH

The nature of performance analysis research

Thomas *et al.* (2005: 5) described a continuum of research, from applied to theoretical, giving three examples of motor learning research of differing levels of relevance to addressing practical problems in sport. They portrayed applied research to be undertaken in real-world settings without experimental control and theoretical research to be undertaken in laboratory settings under carefully controlled conditions. Performance analysis, by its nature, usually involves observational analysis of real-world sports performance. In such situations, it is not possible to either control environmental factors or the quality and tactics of the teams involved. Much of the research done in the discipline is of an applied nature, although there are examples of theoretical research in performance analysis. These theoretical studies typically use supporting evidence from real-world performance data rather than controlled laboratory experiments. For example, the research into Dynamic Systems Theory and how it explains sports performance (McGarry *et al.*, 2002) uses observations from real sports performances in squash and tennis. The recently proposed Interacting Performances Theory uses supporting evidence from tennis matches played at Grand Slam tournaments (O'Donoghue, 2009a). The research undertaken into the effects of rule changes in sport (Williams, 2008) has obvious practical implications. However, the research has also led to an exploratory theory of how players perform in general, given a set of rules and constraints that are imposed upon the match. Motor control theory is relevant to skill acquisition and technique (Strachan and Weir, 2006). This is a theory that can offer explanations for differences in technique between athletes of different levels as well as differences within the same athletes' techniques as a result of training or rehabilitation from injury. Theoretical perspectives of biomechanics include mechanical theory, Newton's laws of gravity, motion, inertia and acceleration due to an acting force, as well as theory from physiology relating to muscle activity and the properties of biological materials.

Sports science is a relatively new area of science where much more work needs to be done in all areas. Performance analysis of sport is an emerging discipline within sports science where even more work is required. Researchers will undertake investigations but when they need to discuss and explain their results, there is often very little sports performance theory upon which to draw. Typically, discussion sections of performance analysis papers will compare findings with those of similar research in the literature, seeking explanations from other disciplines of sports science. For example, time-motion analysis research will draw on evidence from laboratory and field studies in sports physiology to explain results in terms of energy system utilisation. Research into tactics in field games will draw upon coaching literature to explain the findings made. A sustained research effort is required to develop general theories of sport performance and behaviour in games that can also be drawn upon by scholars of performance analysis of sport.

Research, in the broadest sense of the word, is not limited to the activity of university academics and staff within research and development organisations. Police investigations, lawyers researching a case, investigative journalism and market research are all situations where a considerable volume of research is undertaken. These are examples of applied research where the research is being undertaken to fulfil a directly real-world need or address a real-world problem. The continuum of research described by Thomas *et al.* (2005: 5) suggests that research has a range of direct relevance to practical problems. This is certainly true in performance analysis of sport. There is practical study and data collection undertaken in coaching contexts to directly support decisions made by coaches and players in preparation for competition. There are also many academic investigations into sports performance that are not commissioned by coaching needs, but which provide knowledge and practical recommendations that inform practice.

An example of applied research is the work of O'Donoghue *et al.* (2008) to determined percentile norms for performance indicators in British National Superleague netball so that squads can evaluate their performances in matches against different opponents. This is relevant to netball coaching at this level, but it would not be as relevant as research investigations into the particular action that teams have to take in order to achieve higher levels of success. Olsen and Larsen (1997) analysed the performances of the Norwegian national soccer squad allowing progress to be monitored. One of the interesting aspects of the work done by the Norwegian Football Association was that the data gathered were also used to inform theoretical areas, developing a greater understanding of match syntax. One of the best known examples of academic research leading influencing practice is the work of Reep and Benjamin (1968) on the number of passes in possessions that lead to goals being scored. This research showed that more goals were scored from possessions of four passes or fewer than of over four passes.

However, Hughes and Franks (2005) observed that in World Cup soccer there were also more passing sequences of four or fewer passes than of over four passes. Indeed when the proportion of possessions of different numbers of passes that resulted in goals was computed, it was evident that shorter possessions were not the most productive way of producing goals.

Time-motion analysis research can be purely descriptive in nature or can lead to follow-up investigations of sports-specific conditioning programmes. These follow-up studies have a greater relevance to practice than the descriptive studies. Examples of applied time-motion analysis research that include the development and testing of conditioning training programmes include the work of Huey *et al.* (2001), King and O'Donoghue (2003) and O'Donoghue and Cassidy (2002). These research studies either undertook time-motion studies or used existing time-motion results to develop specific training programmes with bursts and recoveries of similar durations to those experienced in real competition. Quasi-experimental studies were then conducted to evaluate the effectiveness of these training programmes for improving relevant fitness test results. As well as training programmes, specific fitness tests can be developed based on time-motion results. Gasston and Simpson (2004) developed a specific fitness test for netball players based on the types of movements performed in the game.

The applied nature of much of the research undertaken in performance analysis of sport does not mean that the research should be any less rigorous than basic theoretical research. Reliability is very important in research located throughout the continuum of research in performance analysis of sport. O'Donoghue and Longville (2004) described their view that reliability is more important in practical performance analysis support work than it is in academic investigations. The reason for this is that important decisions about training and preparation will be made by coaches and athletes based on the performance information produced.

Is performance analysis research? The BASES debate

Although performance analysis has been used to investigate technical, tactical and physical aspects of different sports (Hughes 1993, 1998), there has been a debate as to whether or not notational analysis is research. This debate took place over a few issues of the BASES (British Association of Sport and Exercise Sciences) newsletter. While the debate pre-dated the use of the term 'performance analysis' and was largely focused on notational analysis, some of the criticisms of notational analysis could be directed at performance analysis as a whole and, therefore, it is important to establish that performance analysis of sport is a credible and relevant area of sports science. Hayes (1997a) stated that notational analysis was always done retrospectively and produced historical descriptive results that were unique to the competition being observed. His argument that notational analysis was not research concentrated on the following issues:

- Sports performances cannot be repeated.
- Notational analysis does not follow an experimental design.
- It is not possible to test interventions.

Borrie (1997) responded to these criticisms, arguing that notational analysis could be used for research. First, he stated that research studies in social science and clinical medicine are also done retrospectively. Secondly, he stated that real-time notational analysis systems had been used to support interventions by coaches within competition and a further application of notational analysis is predictive modelling. Borrie (1997) explained that sporting performances do contain unique aspects but also general properties and that good notational analysis research separated these general characteristics of sporting performance from the unique aspects of the data used.

Sharp (1997) entered the debate, describing notational analysis as a research tool for observing, monitoring and collecting data from dynamic and complex situations that could be used in both good and bad research. This is very true and should be recognised by all embarking on performance analysis research whether using biomechanical analysis techniques, qualitative observation or notational analysis.

Hayes (1997b) made a further criticism of notational analysis because it had not been shown to improve performance when applied in a coaching context: 'Show me the results of notational analysis, not the notational analysis results.' Initial attempts have been made to assess the effect of using notational analysis within a coaching context and experiences have been reported (Brown and Hughes, 1995, Jenkins *et al.*, 2007, Martin *et al.*, 2004, Murray *et al.*, 1998). However, much more research is needed in this area.

O'Donoghue (2001b) decided to test the issues of generalisability and repeatability in performance analysis by undertaking a study and then repeating the same study a year later with different data. The intention of this study was that the contribution to the debate about performance analysis research would be the data rather than the author's personal opinion. A computerised notational analysis was done to examine gender and surface effects on singles tennis performance at Grand Slam tournaments. This study used 124 matches from the 1997 French Open, Wimbledon and US Open as well as the 1998 Australian Open. The study was repeated applying the same method for 128 matches from the 1998 French Open, Wimbledon and US Open as well as the 1999 Australian Open. Despite the fact that the two studies used completely different matches where players would have received unique advice from coaches and adopted individual tactics based on relative strengths and weaknesses, the two investigations produced very similar results. This suggested that performance analysis investigations can be done in a reductive way to determine general characteristics of sports performance. This suggested that general characteristics of tennis performance can be described by notational analysis investigations, agreeing with the view expressed by Borrie (1997).

While the results of a performance analysis investigation can be generalised where there are sufficient performances included in the study, the results cannot be taken as representative of the sport beyond the era of the matches that were analysed. Brown and O'Donoghue (2008b) did a 10-year follow-up study of Grand Slam singles tennis using the same system as O'Donoghue used in 1997–9. They found that the nature of tennis has changed in the ten years between 1997 and 2007. While rally durations had decreased in women's singles, they had increased in men's singles. Wimbledon was the only tournament where this trend was not followed, with rally durations increasing in both men's and women's singles. One of the explanations for the changing nature in Grand Slam singles tennis was the introduction of the Type 1 and Type 3 balls at the French Open and Wimbledon respectively in 2002. Such technological advances, developments in player training and rule changes in sports are all reasons why results of performance analysis investigations should not be generalised beyond the era of the data used.

Overlap between performance analysis and other disciplines

There is considerable overlap between performance analysis and other disciplines as most performance analysis exercises investigate aspects of performance that are concerns of other disciplines of sports science. Biomechanical analysis of technique is included within the scope of performance analysis and, therefore, there is an overlap between performance analysis and the wider discipline of biomechanics. Movement analysis results for games are typically discussed in relation to the physiology of intermittent high intensity activity. Analysis of player activity in different match states can be explained with reference to theory developed in sports psychology (Redwood-Brown et al., 2009). Sports psychology and performance analysis overlap in several ways. There are psychological aspects of communication and feedback in coaching contexts that are relevant to studies investigating the effectiveness of performance analysis support. The role of feedback in skill acquisition is also an area where elements of psychology are relevant. There are also psychological aspects of performance itself that can be investigated such as aggression (Shafidazeh, 2008), attentional focus (O'Donoghue and Scully, 1999), arousal, body language and communication. Observational analysis of player activity to assess the injury risk of sports will have an overlap with sports medicine. Talent development is a relatively new area of sports science with its own section of the *Journal of Sports Sciences*. Talent development can be investigated using many complementary methods including analysis of player performances. With video information available for sports events over a period exceeding 50 years, there is a role for empirical observational analysis as well as qualitative observation in sports history research. Sports management and sports development also use performance indicators; for example performance indicators in athletics performance have the potential to be used in funding decisions of governing bodies and high performance directors (Greene, 2008).

HISTORY OF PERFORMANCE ANALYSIS RESEARCH

Performance analysis of sport is a relatively recent discipline of sports science and its history is composed of the history of biomechanics and the history of notational analysis, with these two disciplines coming together within performance analysis in 2001. However, the developments that have led to performance analysis of sport, as it exists today, include the development of sport itself and the history of human movement analysis, which has been influenced by discoveries in anatomy, physiology, mechanics and engineering (Nigg, 2007a). Sport has existed since the ancient Olympics and grew in the 20th century, with the most popular sports being played by professional athletes with increasing media coverage creating sports celebrities even before the Second World War. The changes in sport over the 20th century included increasing importance and increasing reward for success, which has led to greater efforts in preparation. Scientific support for sports preparation is well documented; for example the training manual of the East German athletics federation (Schmolinsky, 1983) pre-dates the establishment of talent development programmes and sports institutes in many countries. The increasing importance and recognition of sport has made it a valuable means of media exposure for commercial sponsors. This has brought money to many sports, which has been used for sports science support as elite performers and teams use any legal means of enhancing performance.

The early history of biomechanics of sport has been well documented elsewhere (Nigg, 2007a, Wilkerson, 1997). In 2007 at the Xth International Symposium of Computer Science in Sport, Gideon Ariel described the progress in biomechanical analysis of sport since 1968. Ariel's analysis of Bob Beamon's world record long jump in the Mexico City Olympic Games of 1968 was the first time that data were gathered from competitive sport for the purpose of biomechanical analysis. The process required to analyse the data was labour- and time-intensive, as Ariel described it in his keynote address to the 6th International Symposium of the International Association of Computer Science in Sport in Calgary in June 2007 (Ariel, 2007). The 16 mm film took three days to be developed before it was manually trimmed to maintain only the frames of interest. Each of these frames was rear-projected onto a matt glass screen so as Ariel could manually measure the lengths and angles of each segment using rulers and protractors. This information was then recorded onto computer keypunch cards for computerised analysis by a mainframe computer allowing kinematic information to be determined. By the time of the 1972 Olympic Games in Munich, Ariel had obtained a sonic digitiser, which speeded up the process of determining joint coordinates and joint angles. Ariel (2007) identified the key developments between 1972 and 1996 as NASA's (National Aeronautical and Space Agency) research to determine the mass of body segments given the height and mass of an athlete, the development of high speed movie cameras with shutter speeds of 200 frames per second, the development of sensitive force platforms, developments in computer software and hardware, portable computers and wireless communication.

At the 1996 Olympic Games in Atlanta, video and other data were uploaded onto the internet for use by coaches, researchers and other interested users (Ariel, 2007). During the 2004 Olympic Games in Athens, further advances in multimedia and communications technology were utilised in the analysis of the discus heats. These advances were:

- miniature cameras that could transmit digital video data to a central computer;
- using the internet to send data to process centres throughout the world;
- wireless control of tripods and cameras from any location in the world;
- automatic digitisation of joint centres;
- automatic report generation suitable for coaches and athletes.

The history of notational analysis of sport has been described by Mike Hughes during keynote addresses of World Congresses of Science and Football (Hughes, 1993) and Science and Racket Sports (Hughes, 1998). Shorthand notations have been used for centuries to record data in many areas of business and science, as well as music. Analysis of movement was done in ancient times using hieroglyphs. The notation of dance has been carried out for centuries in general and rudimentary forms (Hughes and Franks, 1997: 38–9). However, the first recognised system for analysing and recording human movement was Labanotation, founded by Rudolph Laban in 1948. Labanotation enables the recording of many kinds of human motion and is not connected to a singular specific style of dance. In Labanotation, every change made in natural human motion has to be specifically written down (Barbacci, 2002). Due to the need to notate events in real-time and to decrease the complexity of notation, Motif writing was developed (Preston-Dunlop, 1967a, b, c, d) as an alternative to Labanotation, which depicts the key elements of a given movement sequence (Warburton, 2000).

Analysis of behaviour in competitive sport has also been done since the beginning of the 20th century, pre-dating the use of notational analysis. In March 1907, a statistical analysis of the French rugby championship final was published in a newspaper article (Martin, 1907). Hughes and Franks (1997: 40–69) reviewed notational analysis work in different sports occurring as early as the 1970s (tennis, squash and wrestling) and 1980s (volleyball, field hockey, rugby union and Australian Rules football). The key studies undertaken using manual methods were done by Reep and Benjamin (1968) and Reilly and Thomas (1976). Reep and Benjamin (1968) completed a 25-year investigation of the chance of scoring from possessions of different numbers of passes, using data from English league soccer matches played between 1953 and 1968. Reilly and Thomas (1976) used manual methods to estimate distance covered and work-rate during top level soccer performance.

Technological advances were taken advantage of by performance analysts working in academic and practical settings. Video (Lyons, 1988), audio visual aids (Winkler, 1988) and computers (Hughes *et al.*, 1989) were exploited by those studying sports performance. The systems used in computerised notational analysis of sport followed advances in computer storage technology, developments in input and output peripheral devices and greater portability of computer systems (Hughes and Franks, 1995). Early systems that integrated a database of timed match events with the match video used a computer to control video cassette forward-winding and rewinding to display the video sequences of interest (Patrick and McKenna, 1988). Developments in multimedia technology included the storage of video on random access disk. This was exploited by the MAVIS (Match Analysis Video Integrated System) to avoid video tape forward winding and rewinding when accessing the video sequences that satisfied the operator's criteria (O'Donoghue *et al.*, 1995, 1996a, b). Today, commercial systems that integrate video and performance databases are used in competitive sports preparation and by students of university sports science programmes.

In 1992, the First World Congress of Notational Analysis of Sport took place in Burton Manor in England, followed by subsequent World Congresses in Liverpool, England (1994), Antalya, Turkey (1996) and Porto, Portugal (1998). The International Society of Notational Analysis of Sport (ISNA) was founded in 1992 and was responsible for the scientific programmes of these World Congress meetings. It was during his keynote address at the 1998 World Congress, that Keith Lyons proposed the term 'performance analyst' due to the fact that sports performance was by that time being analysed using a broad spectrum of methods not limited to notational analysis. Therefore, in 1999, ISNA was renamed as ISPAS (International Society of Performance Analysis of Sport). In 2001, the importance of computer technology in performance analysis was recognised as the 3rd International Symposium of the International Association of Computer Science and Sport and the 5th World Congress of Performance Analysis of sport were combined within a joint conference in Cardiff, Wales called PASS.COM (Performance Analysis, Sports Science and Computers). The increased volumes of research and development in both areas allowed these two series of conferences to proceed individually from 2004 with the 6th, 7th and 8th World Congresses of Performance Analysis of Sport taking place in Belfast, Northern Ireland in 2004, Szombathely, Hungary in 2006 and Magdeburg, Germany in 2008.

The volume of research being undertaken has increased so much that the two-year interval between World Congresses is considered by the ISPAS to be too long to keep pace with research and developments in the area.

Therefore, a series of International Workshops in Performance Analysis of Sport commenced in 2007 providing an annual forum for the communication of ideas in addition to the bi-annual World Congresses of Performance Analysis of Sport. The first International Workshop was held in Cardiff, Wales in 2007 and emphasised developments in commercial performance analysis systems. The second International Workshop was hosted by Leeds Carnegie University in 2008 and emphasised performance analysis in coaching, particularly in Olympic sports. The University of Lincoln hosted the 3rd International Workshop in 2009, which was a forum for academic research.

ISPAS founded the *International Journal of Performance Analysis of Sport* in 2001 and the first volume published the keynote papers from the PASS.COM conference. In 2008, the journal published three issues of its eighth volume with a total of 40 research papers covering a range of sports, written by authors from 17 different countries.

RESEARCH TOPICS IN PERFORMANCE ANALYSIS

Critical incidents and perturbations

Performance analysis can involve gathering data from hundreds if not thousands of events that occur during games. However, it is often very difficult to uncover patterns that clearly distinguish between successful and unsuccessful performances. The outcome of a match can be determined by a small number of critical incidents and focusing on them can lead to more efficient and effective analysis of sports performance. In squash, a rally can be considered to be in a stable state with an observable rhythm of behaviour as the two players alternately move away from the 'T' position to play a shot and then return to the 'T' position in readiness to travel to wherever the next shot needs to be played from (McGarry *et al.*, 1999). A 'perturbation' is an event that disturbs this rhythm and leads to the rally becoming unstable. This is usually to one player's advantage and the other player's disadvantage. From this point, the rally can either remain unstable until it ends with the perturbation having played a clear role in the outcome of the rally, or the rally can gradually return to a state of stability. A rally may contain more than one perturbation and alternate between stable and unstable states.

In team games, the concept of perturbations has allowed researchers to abstract sports performance to the critical incidents that occur (Hughes *et al.*, 2001a). Given the impact that critical incidents and perturbations have on the outcome of a match, there is a clear rationale for teams attempting to avoid such incidents that place them in a vulnerable position within a match. Very often in sport, there is a series of events that can be identified after the match that lead to a score being conceded. This tracing backwards from the score being conceded can reveal a path from a root cause to the critical situation and subsequent score. This is very similar to the way safety requirements of software-intensive systems are analysed (Ericson, 2005), except that the analysis is done during system development to help avoid hazardous conditions and possible mishaps. O'Donoghue (2007c) described how fault tree analysis could be used to identify critical states from where the path followed could lead to a vulnerable situation or an alternative path could lead to a safe state. A key difference between the use of fault trees in safety analysis and their use in performance analysis of sport is that in sport, a team will wish to create perturbations that they can take advantage of.

Analysis of coach behaviour

Analysis of coach behaviour has become an established area of research (Gilbert and Trudel, 2004, Kahan, 1999, More, 2008, Potrac *et al.*, 2002, Van der Mars, 1989). There are different quantitative methods of analysing coach behaviour, including the Computer Aided Coaching Analysis Instrument (Franks *et al.*, 2001, Harries-Jenkins and Hughes, 1995, More and Franks, 1996), the Revised Coaching Behaviour Recording Form (Côté *et al.*, 1995, Durand-Bush, 1996) and the Arizona State University Observation Instrument (ASUOI) (Lacy and Darst, 1984, 1985, 1989).

There are many aspects of coaching such as the knowledge of the coach, the decisions made by coaches and the development of training programmes that are best studied using methods other than performance analysis. However, coaching and teaching style as well as coach behaviour during coaching sessions and competition can be observed and analysed in detail. The behaviour of high level and successful coaches has been reported (Bloom *et al.*, 1999, Gilbert and Jackson, 2006). The ASUOI is becoming a recognised standard for the analysis of coach behaviour and it has been used to analyse the behaviour of strength and conditioning coaches (Massey *et al.*, 2002), coach behaviour during ice-hockey games (Trudel *et al.*, 1996), age group effects (Cushion and Jones, 2001) and the behaviour of physical education teachers (Paisey and O'Donoghue, 2008).

Developments in computer technology have not only led to more sophisticated match analysis systems but also to more sophisticated systems for analysing coach behaviour. For example, Brown and O'Donoghue (2008a) used a split screen system to allow simultaneous viewing of the coach and the wider training session. The use of microphones that can be worn by the coach with sound being transmitted to a receiver connected to a camcorder allow the words of the coach to be recorded in detail while the camera may remain distant enough not to interfere with the training. Qualitative analysis can complement the quantitative analysis of coach behaviour, providing explanations of the behaviour that is recorded. Paisey and O'Donoghue (2008) analysed physical education teachers' behaviour using the ASUOI as well as in-depth qualitative observational analysis of the video recordings of the physical education lessons. Donnelly and O'Donoghue (2008) analysed netball coach behaviour at three different levels using the ASUOI, and used a follow-up interview to discuss the similarities and differences in the behaviours of the coaches.

Performance indicators for different sports

Performance analysis allows the complex and dynamic nature of sports performance to be represented in an abstract way, using performance indicators that focus attention on the most relevant characteristics. The term 'performance indicator' is not another name for 'variable' but is a term for those variables that are demonstrated to be valid measures of important aspects of performance and which possess the metric properties of having an objective measurement procedure, a known scale of measurement and a valid means of interpretation. Aiming for these qualities for performance indicators in performance analysis will help ensure that the term 'performance indicator' is used in a similar manner to the way it is used in business and engineering fields.

Different investigations have used differing sets of variables to characterise sports performance. For example, some time-motion analysis investigations look at the distribution of time among different classifications of movement (Bangsbo et al., 1991, Bloomfield et al., 2004, O'Donoghue et al., 2005b). Other time-motion investigations use indicators relating to distances covered and velocity profiles (Di Salvo et al., 2009, Hughes et al., 1989, Reilly and Thomas, 1976, Withers et al., 1982). Similarly, studies of tennis strategy fail to agree on standard sets of variables to use to represent strategy. For example, Hughes and Clarke (1995) used player positioning, ball placement and rally times. O'Donoghue and Ingram (2001) used point types, especially the percentage of points where players attack the net. Therefore, since the advent of the term 'performance indicators' in performance analysis of sport (Hughes and Bartlett, 2002), there has been a research effort into defining the most valid performance indicators in different types of sports. Hughes and Bartlett (2002) classified formal games as invasion games, net games, wall games and striking/fielding games and defined the types of performance indicator of interest when analysing performances in those sports. There are other sports that do not fall within the classification made by Hughes and Bartlett (2002) where performance indicators have been proposed. For example, canoe slalom (Wells et al., 2009), middle distance athletics (Brown, 2005) and martial arts (Shapie et al., 2008) are sports where performance indicators have been proposed based on the broad principles outlined by Hughes and Bartlett (2002).

Varying approaches have been proposed to identify the key performance indicators to characterise different aspects of sports performance. Focus groups have been established to obtain expert opinion as to the key aspects of matches to record and present (McCorry et al., 1996). Artificial neural networks and regression analysis have been used to identify the factors most associated with outcome indicators in tennis (Choi *et al.*, 2006b). Tennis is an interesting sport to consider, as five set matches can contain periods where the eventual winning player is losing. Furthermore, in real-time feedback systems that can be used within the match, it is necessary to use performance indicators based on sections of the match. Therefore, Choi *et al.* (2008) used individual quarters of basketball matches and individual sets of tennis matches to determine the performance indicators most associated with winning performance. Statistical analysis has also been undertaken to compare the winning and losing teams within matches to identify the performance indicators that distinguish between them (Choi *et al.*, 2006a). This approach can be criticised because some matches will be played between very successful teams. Therefore, an alternative approach is to classify teams according to finishing positions within tournaments and identify performance indicators that distinguish between successful and unsuccessful teams within tournaments (O'Donoghue *et al.*, 2008).

Investigations comparing winning and losing teams within matches and successful and unsuccessful teams within tournaments can identify many performance indicators as being significantly different between the samples of teams being compared. One of the disadvantages of such investigations is that many of the matches used are between teams or players with a large difference in ability. Such matches, where the outcomes are wins for the higher ranked of the two teams (or players) may not be the most critical ones for coaches to prepare for. The most even matches, where the result could be a win or a loss with almost equal probability are the most important matches. Therefore, recent research has attempted to determine the performance indicators that distinguish between winning and losing performances in matches between closely ranked teams (Csataljay *et al.*, 2008).

Very often, the performance indicators identified though peer review or more quantitative methods will contain pairs of performance indicators that are highly correlated to each other. A group of correlated performance indicators represent the same broad aspect of performance or at least different aspects of performance that are strongly associated. Therefore, O'Donoghue (2008a) proposed the use of principal components analysis to identify independent broad dimensions within sports performance and those performance indicators associated with them. The performance indicator most highly loaded onto a principal component could be selected to represent the factor of interest.

The biggest problem with many of the quantitative techniques used to identify valid performance indicators is that they require a lot of data to already exist. The purpose of identifying valid performance indicators may be to develop a system to allow these to be analysed, which is a 'chicken and egg' situation. We need the system first to gather the data to test the validity of the performance indicators, but we need to know the performance indicators first in order to develop the system! However, there may be internet sources of data that are available for such exercises, or the purpose of the entire study may be to identify valid performance indicators for a given sport that can then be utilised in practice by others.

Work-rate analysis and evaluation of injury risk

Analysis of work-rate has been undertaken using both manual and computerised methods to determine distances covered and the breakdown of match time among different movement classes (O'Donoghue, 2008b). Speed, agility and quickness training programmes are being used to prepare athletes in many sports and yet there are very few studies giving an understanding of the agility requirements of competing in those sports. This type of research is very time-consuming even when computerised systems are used for data entry (Bloomfield *et al.*, 2007a, b, Robinson and O'Donoghue, 2008). However, the results of such research are of clear benefit to those developing the conditioning elements of players' training programmes.

The technique developed by Bloomfield *et al.* (2004) has not only been used for the analysis of agility requirements of soccer (Bloomfield *et al.*, 2007b) but also for the analysis of injury risk in netball (Darnell *et al.*, 2008, Williams and O'Donoghue, 2005). Other research analysing injury risk has used observational analysis of match events (Hawkins and Fuller, 1998, Robinson and O'Donoghue, 2008), with some work also classifying the risk of injury associated with each event (Rahnama *et al.*, 2002). Further work is needed in many other sports to assess their potential for injury.

Reliability of methods

The reliability techniques used in performance analysis can be challenged for failing to detect poor reliability and also for falsely concluding that reliable methods are unreliable. The main problem with the reliability methods that have been used is that there have been very few attempts to relate the reliability statistics to the analytical goals of the studies for which the systems are used. Very often a level of inter-operator agreement is set as a maximum percentage error of 5 or 10 per cent. However, there is no rationale given for this and the impact of a 5 or 10 per cent error on the investigation being conducted is unknown. Choi *et al.* (2007) synthetically introduced different severities of error into basketball data to determine the kappa values that would be associated with these. By determining the kappa values (or any other reliability statistic) that we would get for an acceptable level of error or the point at which errors would lead to the incorrect conclusion being drawn about the performance, we can identify threshold values for the reliability statistics that are more meaningful than the arbitrary values that have been used to date. O'Donoghue (2007b) showed that some reliability statistics did not have construct validity in that values were obtained when comparing completely different performances that exceeded those obtained when the same performance had been analysed twice by an expert observer. Therefore, there is a great deal of work that needs to be done urgently to produce reliability assessment procedures that are themselves valid assessments of system reliability in relation to the analytical goals of the studies for which they are to be used.

Analysis of technique

Detailed analysis of technique remains an important area of research in performance analysis of sport. Lees (2008) described 'event skills' as skills that are themselves sports events, such as the long jump. Optimal techniques for maximising performance outcomes of event skills is an area that needs to be applied to different sports, with improving technology allowing more efficient data gathering which, in turn, allows more subjects to be included in studies. Lees (2008) described 'minor skills' as skills that are performed repeatedly during competition, for example the service in tennis. There may be different types of serve based on pace, placement and the application of spin. Repeated skills in games such as tennis cannot apply a single optimal technique as the opponent will be able to anticipate the service. There are very few sports, if any, where every variation of each minor skill has been analysed for players at all ability levels. Therefore, there is a need for further research in this area.

'Major skills' are performed repeatedly but have a much greater impact on success in a sport than minor skills (Lees, 2008); these include the golf swing. When one considers the golf swing, there are a variety of techniques based on the distance a shot has to be played, among other factors. The variety of applications of major skills in many sports gives rise to a great deal of opportunities for original research.

Changing equipment and regulations in sport can render previous research into technique obsolete and so up-to-date research is always needed. Variability in sports performance is an emerging research area, with biomechanics investigations showing variability in successive strides when running (Heiderscheit et al., 1998) and walking (Dingwell et al., 1999).

Technical effectiveness

Technical analysis is concerned with how well skills are performed in sport. Positive to negative ratios have been used to evaluate the skills of players in soccer (Gerisch and Reichelt, 1993, Olsen and Larsen, 1997, Rowlinson and O'Donoghue, 2009), while winner-to-error ratios have been used in racket sports (Murray and Hughes, 2001). A criticism of the use of positive-to-negative ratios is that all events must be judged as wholly positive or wholly negative. One only has to look at a soccer match any weekend to see within the first 10 minutes of play a one-to-one situation where neither player fully achieved their objective. Another issue is that the degree of difficulty of the situation where the skill is performed is often not taken into consideration. Hughes and Probert (2006) used a quality rating of -3 to +3 to give greater precision to technical effectiveness in soccer. Much more work of this nature needs to be applied to many other sports. The effectiveness of players of different standards and genders, when performing in different situations, is still a key research area.

Tactical patterns of play

Tactical analysis is one of the main purposes of performance analysis of sport and there are many sports where much more research is needed into patterns of play. Indeed, when one thinks of a particular sport such as tennis, there are a multitude of different tactical aspects such as service tactics (Unierzyski and Wieczorek, 2004), tactics on return of serve, tactics in different game states (O'Donoghue, 2006a, 2007a, Scully and O'Donoghue, 1999), tactics on different court surfaces (O'Donoghue and Ingram, 2001) and opposition effects on tactics utilised (O'Donoghue, 2009b). Sports are played at a range of levels, from recreational to international and world class performance, and there are no sports where all aspects of tactics have been analysed at all levels of the sport. The tactics of winning and losing teams within matches is also a useful area of research but does suffer from the disadvantage that both teams in some matches might be very successful teams; for example world cup finalists in cricket, rugby and soccer. Therefore, an alternative approach is to analyse the tactics of successful teams and unsuccessful teams within tournaments based on finishing position (Taylor et al., 2008). For example, the quarter finalists of the soccer world cup could be compared with those teams that were eliminated after the group stage. The effects of gender (O'Donoghue and Ingram, 2001), venue (Taylor et al., 2008) and rule changes (Williams, 2008) are other factors that may have an influence on tactics.

Performance profiling

Performance indicators in sports performance are not stable variables like anthropometric variables, and there are many sources of variability in sports performance including opposition effects, venue and score line (Taylor et al., 2008). Therefore, teams and individuals need to be represented by performances from multiple matches. A performance profile in performance analysis is a collection of performance indicators that together characterise the typical performance. Hughes et al. (2001b) produced the first technique for analysing multiple performances, but it did not produce a profile that incorporated all of the performance indicators of interest into a single profile. Instead, each performance indicator was dealt with in isolation to determine the number of matches required for the accumulating mean to fall within a tolerable level of error of the overall sample mean. O'Donoghue

(2005a) produced a profiling technique that displayed the various perform-ance indicators of interest on a radar chart profile that related the typical performance to the norms for the relevant population of performers. This technique also showed the spread of performances about the typical per-formance, allowing consistent and erratic areas of performance to be repre-sented. James *et al.* (2005) used a form chart to show the 95 per cent confidence limits of performance indicators, allowing different groups of performers to be compared. These techniques can be used to produce pro-files of different levels of performance in different sports. The study of vari-ability in sports performance is still a major research topic that can also be investigated through the use of performance profiling techniques.

While profiling is important in understanding the performances of an opponent, in scientific research the use of profiling maybe a hindrance. Where there are enough players involved in an investigation, it is not necessary to have each player represented by multiple performances. Where a single per-formance is used for each player, the performances of some players will underestimate the typical values of some performance indicators while others will have overestimates of the typical values of performance indicators. The additional variability caused by individual match effects actually means that the researcher can be more confident in any significant results found. A pro-gramme of research has commenced into the impact of limited sample sizes, individual performances and limited reliability in performance analysis of sport (Ponting and O'Donoghue, 2009). This programme of research will determine the effects of such factors on the conclusions drawn from studies. This research uses a fictitious population of performances from a fictitious sport called 'pseudoball'. This allows the conclusions drawn from analyses of samples to be compared with the known truth based on the synthetically created population. Further research is needed in this area to understand where it is necessary to use multiple match profiles for performers and where individual performances will be sufficient.

The effectiveness of performance analysis support

There is still scepticism about the effectiveness of performance analysis in enhancing sports performance. Studies to analyse the effectiveness of such support are impossible to control but are none-the-less very important. So far studies in field hockey (Boddington, 2002), squash (Brown and Hughes, 1995, Murray *et al.*, 1998), Gaelic football (Martin *et al.*, 2004) and netball (Jenkins *et al.*, 2007, Mayes *et al.*, 2009) have provided mixed evidence on the effectiveness of performance analysis support. Where teams and players do improve, it is impossible to prove that it is as a direct result of the use of performance analysis. This has deterred many from undertaking such inves-tigations. However, the research is vitally important and hopefully over the next few years many studies of this type will be done so that the balance of evidence from all published studies can be used to demonstrate the effectiveness of performance analysis support.

Analysis of referees and officials

Players in individual and team sports train hard to perform in competition and, therefore, expect that the competition will be officiated fairly and competently. Indeed, there have been many occasions where the performance of a referee has been a large factor in determining the result of a match. Therefore, the performance of referees and officials is an important area of investigation (Hartshorn, 2009). There are sports where the rules are specified rigorously, suggesting that violations of the rules can be observed. However, there have been many occasions where commentators and broadcast panellists analysing the same video recording of an incident will draw completely different conclusions about the action the referee should have taken. This can be due to different interpretations of the regulations, the limitations of two-dimensional video images and player deception. Mellick (2005) investigated how referees communicated their decisions to players and made recommendations for rugby union referees. Referee positioning is important to be able to obtain a clear view of incidents and make appropriate decisions. A team can use teamwork to be able to achieve their objectives, and not all players have be directly involved in on-the-ball play at all times. It is necessary, however, for officials to be able to follow play closely enough to be able to observe player behaviours that could be in violation of the game's regulations. Therefore, the fitness of referees is tested in some sports, and training programmes for referees have been developed. Analysis of referee work-rate and movement is important to be able to inform the development of such tests and training programmes with a fuller understanding of the physical demands of refereeing a match. Time-motion investigations of referee performance has been done in soccer (D'Ottavio and Castagna, 2001), Gaelic football (Gamble *et al.*, 2007) and rugby union (Hughes and Blunt, 2001, Mizohata *et al.*, 2009).

SUMMARY

Performance analysis is a discipline of sports science that overlaps with physiology, coaching science, psychology, talent identification and sports medicine due to the fact that performance analysis investigations analyse some aspect of performance, whether it is physical, tactical, technical or behavioural. The analysis of a large enough sample of performance allows general characteristics of performance to be determined rather than the research being a retrospective measurement of specific individual performances. Performance analysis research is largely applied research, although some basic theoretical research has been done. The methods used in performance analysis continue to increase in sophistication, facilitating higher quality of research with greater volumes and accuracy of data. There is a large array of research topics in performance analysis, all of which provide opportunities for interesting and important research.

CHAPTER 2

RESEARCH PROCESSES

INTRODUCTION

This chapter is intentionally entitled 'research processes' to make the point that there are many different ways of undertaking performance analysis research. There are common activities to all of these different processes such as the need to select a viable and worthwhile research question and explain findings with reference to theory. However, there are also specific tasks associated with qualitative research, quantitative research and mixed methods that are discussed in this chapter. Some research processes in performance analysis follow a fixed design with formal hypotheses to be tested. Other research processes permit much greater flexibility to follow emerging lines of enquiry. This chapter will classify the main types of research processes in performance analysis of sport. The most common types of research used in performance analysis are descriptive retrospective research, case studies, quasi-experimental studies and observational analysis.

In this chapter, a distinction is made between types of research and data gathering and analysis techniques. When describing types of research, broad umbrella techniques are covered that could use individual data gathering and analysis techniques. For example, a case study is an N=1 design that could be a quantitative performance profile, an account or an interview. Similarly, historical research is used to study events in past eras, but this could be done using a variety of different data gathering and analysis techniques. Developmental research is a type of research that could involve the use of questionnaire surveys or more invasive testing and measurement during data gathering. In this chapter, types of research are considered separately from methods used to gather and analyse data. The underlying assumptions of research methods are described before quantitative, qualitative and mixed method research processes are covered.

TYPES OF RESEARCH

Observational research

The main type of research used in performance analysis of sport is observational research. In particular, notational analysis studies involve empirical observation with events being counted and timed. Observational research is distinguished from experimental research in terms of the trade-off between control and ecological validity. Experimental research typically involves participants engaging in activity under controlled conditions, sometimes in a laboratory setting, so that the effect of manipulated independent variables on hypothesised dependent variables can be investigated. Observational research does not offer this level of experimental control but has a higher level of ecological validity. Ecological validity represents how well the conditions of a study reflect the real world. In observational analysis techniques used in performance analysis of sport, it is common for sports performance in real competitions to be studied. Other quantitative data that can be recorded during observational research include distances covered, speeds of movement and locations of events.

This book considers types of research differently to other textbooks such as that of Thomas *et al.* (2005). One difference is that Thomas *et al.* (2005: 19) considered job analysis and observational research to be different types of descriptive research. This book views job analysis as an application of other types of research including observational research. Another difference is that in Thomas *et al.*'s (2005) textbook, qualitative research is considered to be a separate type of research to observational research. In the current book, qualitative methods are seen as methods of gathering and analysing data that can be used within different types of research including observational research. Participant observation is very valuable as the researcher learns from experiencing the situations of interest. However, in performance analysis research, non-participant observation is more appropriate as the researcher may not qualify as a participant for the investigation and it is important to remove any sources of bias from the data used. The commonly held view that notational analysis is a highly quantitative type of research is not entirely true. When one considers the activity done during notational analysis, human behaviour is observed and subjectively classified according to a set of behaviour categories being used. For example, in time-motion analysis, there is a subjective judgement as to whether a player is jogging or running. Therefore, the counts and timings of movements are simply quantitative counts and timings of the qualitative judgements that have been made by the observer. Some research in performance analysis has very little quantitative analysis. For instance, research into tennis player gaze when playing shots can be done through expert analysis of photographs without using any numerical measurements (Lafont, 2007, 2008).

Experimental research

As has already been mentioned, experimental research involves participants engaging in activity under controlled conditions for the purpose of investigating the effect of the experimental condition on some hypothesised dependent variables of interest. Performance analysis of sport essentially involves the analysis of actual performance in competition or training. Even when laboratory-based studies of key technical skills are undertaken in performance analysis, the investigations are highly quantitative descriptive studies based on detailed measurements rather than experiments into the effect of some independent factor. Therefore, performance analysis of sport does not include experimental research. However, there are many examples of research investigations that are comprised of descriptive observational parts and experimental parts. For example, Huey et al. (2001) used time-motion analysis to investigate the demands of field hockey competition before developing a specific intermittent high intensity training programme and testing this programme using a quasi-experimental study. When explaining performance analysis results, it is often necessary to draw on evidence from experimental studies in other disciplines such as physiology, psychology and biomechanics. As a consequence, it is important for performance analysts to have a good understanding of experimental research and other types of research in sports science.

There are many different experimental designs that are distinguished by whether there are control as well as experimental treatments, whether there are pre-tests and mid-tests as well as post-tests, placebos, blind testing, random assignment of participants to groups, whether participants are assigned to groups before or after testing and whether participants are exposed to one or more experimental condition. Thomas et al. (2005: 330–43) listed no fewer than 13 different experimental designs within three broad classes of experiment; pre-experimental, true-experimental or quasi-experimental studies. This list is by no means exhaustive and variations on the different designs are possible.

The main true-experimental design described by Thomas et al. (2005: 333–5) is the 'pre-test – post-test randomised-groups design', where two independent groups of participants are tested before and after some experimental period; one group will have been exposed to an experimental treatment condition while the other will be under control conditions. The advantage of using a control group is that if there is a change in the experimental group's test performances over the experimental period, this can be compared to what happens over that period for participants who are not exposed to the experimental treatment. The use of pre-testing as well as post-testing is important because without pre-testing we cannot be sure whether differences between the groups' tests performances were developed during the experimental period or whether they already existed before the study. Scientific evidence of an effect of the treatment on the hypothesised dependent variables being tested would require control of all aspects of participants' lives (eating, drinking, sleeping, activity and so on) with the exception of the experimental treatment, which would be manipulated by the researcher during the experimental period. However, it is neither ethical nor feasible to control the lives of participants to this extent. Therefore, true-experimental designs use random assignment of participants to the control and experimental groups. If there is a large enough number of participants in the experiment and they are randomly allocated to groups, it is hoped that the variation in sleep patterns, diets, physical activity level and other lifestyle variables that may influence test performance will be similar between the two groups of participants.

Pre-experimental designs do not have random assignment of participants to groups and do not involve a control group. One pre-experimental design described by Thomas *et al.* (2005: 331) is the 'static group comparison', where an experimental group is tested after an experimental period. The post-test results are compared with post-tests results from participants who were not exposed to the experimental treatment. However, in this design participants are not grouped for the purpose of the experiment: the groups already existed prior to the experiment. For example, a group of participants who engage in a certain type of training may be compared with a group formed from those who do not do such training.

Ex-post-facto designs are quasi-experimental designs that involve participants being tested but then grouped after the experimental period has been completed. A good example of this is McLaughlin's (2003) PhD study, which involved time-motion analysis of 32 primary school children on seven occasions over a two-year period. At the end of the period, the children's absenteeism and body mass indexes were used to divide them into two groups: the healthiest 16 children and the least healthy 16 children. These groups were compared in terms of the activity observed in the playground at break times and lunch times to determine if there was an association between voluntary involvement in moderate to high intensity activity and health. This study did not qualify as a true experiment for two reasons. First, the groups were not assigned randomly and there was no experimental treatment. Secondly, observations of movement behaviour were used, although standard tests and measurements were also used within other studies that were included in the PhD.

Another type of quasi-experimental design is the 'pre-test – post-test groups design', which is the same as the 'pre-test – post-test randomised groups design' except that participants are not randomly assigned to the experimental and control groups. It is often not feasible to randomly assign participants to groups as some may be prepared to perform the pre- and post-tests, but are not willing to undergo the experimental treatment. For example, O'Donoghue and Ormsby (2002) devised a test of free taking performance in Gaelic football that they used in an experiment into the effect of mental imagery training. Some participants were able to perform the pre-test and post-test but did not wish to do the relaxation training two days for six weeks. An issue with this study is that there was a possibility that those participating in the experimental group had a different attitude towards psychological skills training in general than those who were members of the control group. Therefore, there may have been psychological effects influencing the results of the post-tests. Pre-experimental designs and quasi-experimental designs can be used but their disadvantages compared to those of true-experimental deigns must be recognised.

A further type of experimental design is the cross-over design where all participants are tested following both experimental and control conditions. Because participants may perform the test better the more they do it, the participants are divided into two groups that are exposed to both the experimental and control conditions. One group is exposed to the experimental condition first and control condition second, while the other group is exposed to these conditions in the reverse order. One of the main advantages of the cross-over design is that inter-individual differences are eliminated from the study as the effect of both experimental and control conditions are tested using the same sample of participants.

There are two main types of validity of experimental studies: internal and external validity. Internal validity is the extent to which any changes in the test performance can be attributed to the experimental treatment. Observed changes may be due to participants maturing over the experimental treatment, familiarisation with the testing process, biased allocation of participants to groups and psychological expectancy factors. External validity is the extent to which the results of the experiment can be generalised. External validity requires participants who are representative of the population of interest to be used as well as tests that have ecological validity.

Historical research

Historical research is important as solutions to contemporary problems can be based on past lessons. The importance of historical research is recognised in sports science with several journals publishing sports history research, including *Sport History Review, Journal of Sport History, International Journal of the History of Sport, Canadian Journal of the History of Sport, Western Historical Quarterly* and *British Journal of Sports History*. The sources of data used in historical research are classified as primary and secondary sources. Primary sources include biographies, newspapers, minutes of meetings, memoirs, other documents and video recordings that are directly related to the events of interest. It is sometimes possible to use participants who were involved in the historical events of interest; these individuals are primary sources of data. Secondary sources of data are indirectly related to the events of interest and include quoted material, reproduced material and published research that has investigated the events of interest. People who lived during the time of the events of interest but who were not directly involved in the events can also participate in historical research as secondary sources of data.

An issue in historical research is the need to establish the authenticity of the data used. Documents and artefacts may be genuine or they may be hoaxes or other misinformation. Therefore, in published papers in sports history, the accompanying reference list and notes state whether documents such as minutes of meetings are signed or not. Controversial social and political issues have an increased likelihood of being portrayed by biased viewers. Therefore, the context of any data used needs to be understood when establishing the reliability of the data. Performance analysis research techniques can be applied within historical research. For example, changes in Grand Slam tennis since the 1960s can be studied using video recordings of sets of matches from different decades. Such a study may address a need by the media. For example, one might wish to produce a television documentary programme about changes in the game over a period of 50 years to be broadcast in the build up to a Grand Slam tournament.

Developmental research

Developmental research is concerned with changes in variables over a long period of time such as the human lifespan or a career in sport. There are two types of developmental research: longitudinal studies and cross-sectional studies.

Longitudinal studies investigate changes in the same participants over a period of time. For example, motor abilities could be tested over various points of human development from childhood to adulthood. Performance variables within fitness tests or during competition could also be monitored over a long period of time. Talent identification is a broad multi-disciplinary area of sports science research where performance analysis has a role to play within longitudinal studies to try to identify areas of performance that are associated with successful adult performance and the times within athletes' careers where these can be identified. Prospective longitudinal studies monitor participants who can be grouped according to some hypothesised talent factors at the beginning of the research. An example of this was a survey of young Belgian soccer players' practice over a period of 18 years (Helsen *et al.*, 2000). Retrospective longitudinal studies monitor participants, but do not group them for the purpose of analysis until later years of the study where they can be distinguished according to some status or criteria of interest. Cross-sectional studies, on the other hand, use different participants to monitor changing trends in variables between cohorts. For example, ProZone® (2007) has reported trends in movement during FA Premier League soccer over successive seasons where there will have been some changes in players participating in the competition.

Survey research

Survey research describes the reportable characteristics, beliefs, attitudes, opinions and intentions of populations of interest using samples. Occasionally censuses are done to survey the whole population, but usually a sample is used in research. Surveys can either be done in an exploratory way, with few theoretical assumptions, or can be used to test some pre-existing theory. Survey research often involves self-reports such as interviews and question- naires, but occasionally surveys can be conducted by researchers gathering information without asking participants to be interviewed or complete questionnaires. Such surveys include comparing prices between retail outlets or surveying road traffic. What distinguishes these types of surveys from observational research is that observational research involves detailed anal- ysis of behaviour while price or traffic surveys would record data other than behaviour. Fitness surveys do not involve observation and analysis of behav- iour in detail, but instead record key results of tests and measurements made on participants. An example of a fitness survey is the 1989 Northern Ireland Fitness Survey, which produced norms for Northern Ireland children for anthropometric measurements, fitness test results and reported lifestyle var- iables (NIFS, 1989).

Undertaking a survey requires a lot of planning and pilot work to ensure that the surveying technique being used is clear, complete and understand- able. If interviewers are being used, they must be trained to interview par- ticipants for the purpose of the study. If self-completion questionnaires are being used, the questionnaires and accompanying instructions must be thor- oughly checked for clarity and completeness. If a large batch of question- naires containing mistakes is sent out to participants, the exercise will have to be repeated at cost, and some participants will not respond if they see the research as being done incompetently.

Surveys are also being done on the internet with specifically designed data capture forms and individual responses being stored in databases for efficient processing and analysis. An important issue with internet surveys is that there is some means of ensuring the data are authentic. This could be done using e-mail addresses of potential participants that may be publicly shown on organisation internet sites. Telephone surveys are also possible, and have the additional advantage over questionnaires that responses can be clarified and some probing into answers can be done.

Performance analysis can sometimes be used in conjunction with survey research, where the investigation requires both observational and self-report data. A good example of this was a study of scoreline effects on mood and work-rate of semi-professional soccer players (McStravick and O'Donoghue, 2001). This is a particularly good example as the data were collected and analysed within a level 3 research project. The student had been critical of previous research into the effect of scoreline on work-rate because the performance analysis methods used left a large pathway of mechanisms that may be involved as a 'black box'. The previous study by O'Donoghue and Tenga (2001) showed a significant scoreline effect on work-rate, but had to speculate on the processes that may have been responsible. Therefore, McStravick combined the use of performance analysis with survey techniques in order to gather work-rate data and mood data respectively. The work-rate data were gathered during observation of player performances during Northern Ireland Premier League soccer matches. After the match was completed, McStravick met with the player and recorded mood ratings reported by the player during parts of the match where the player's team was in different scoreline states.

Action research

Research improves our knowledge of phenomena of interest and provides evidence that can be used by policy makers, coaches or other people. The decisions that are taken often relate to practice and the benefits of changing practice should be supported by evidence. Therefore, research and action are often done separately by different groups of individuals: researchers and policy makers. Some research findings may never be used to inform decisions, while other research studies may be commissioned by policy makers to obtain independent evidence to inform decisions. The term 'action research' represents a cycle of action and research that links these two activities more closely. In action research, an act of intervention is taken within a real-world situation and the effect is closely monitored. Action research can be done on a range of scales from the reflective practice of individual professionals (Schön, 1983) to a very large scale for entire organisation change (Zuber-Skerritt, 1996).

Action research is appropriate whenever specific knowledge is required about a practical problem. The role of performance analysis within a coaching context is essentially an action research role that follows four key stages of action research: observing, reflecting, planning and action. Figure 2.1 shows a modified version of a model of performance analysis activity within the coaching process (Mayes et al., 2009). When a match is played, it is observed with match events being entered into a match analysis system, live. Match statistics are produced once the match has completed, although some live analysis systems also have the capability of producing match statistics during the match. Performance indicators are compared to norms for the level of opposition faced in the match (O'Donoghue et al., 2008) and areas of play requiring attention are identified. Commercial match analysis packages such as SportsCode (SportsCode, Warriewood, NSW, Australia), Focus X2 (Elite sports Analysis, Delgaty Bay, Scotland) and Dartfish (Fribourg, Switzerland) have password-protected internet sites for relevant video sequences to be uploaded and viewed by the squad. This helps the squad reflect on the performance prior to the next training session. Coaches can

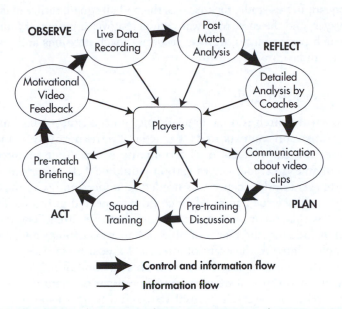

Figure 2.1 Performance analysis in a coaching context (Mayes *et al.*, 2009)

plan training sessions to specifically address areas where the squad need to improve. There are often several areas requiring attention and coaches need to prioritise these in terms of their importance to the match outcome and how difficult they are to remedy. These can be discussed with the squad prior to training drills commencing. Planning is also done prior to the next match using information the squad may have about their next opponents. The action taken in training can then be evaluated when the performance in the next match is compared with previous matches played. This activity is repeated in a cyclic fashion throughout the season in a process of gradually improving the quality of performance.

Case studies

A case study probes deeply into the case of a single unit (individual, group or organisation), using quantitative and/or qualitative methods. Deep analysis of the particular case is essential for the project to be justified as a level 3 or level 4 dissertation. A time-motion analysis of a single player performance using methods where the data could be captured during live observation of the match would not be sufficient for a research project. If a student were particularly interested in studying the performance of an elite performer as a case study, this could be done using a large enough series of matches. The methods could also be expanded so as multiple data sources were being used and so that the player was being investigated in sufficient depth. Jenkins *et al.* (2007) used a case study approach to investigate the effectiveness of performance analysis support to a netball squad. In this study, the case was the netball squad, rather than an individual player, and the case study was conducted over a period of four months where match performances, training sessions, coach decisions and player attitudes were studied.

Ex post facto research

Ex post facto research is done in academic, journalism, medical as well as police investigation contexts, and is carried out retrospectively in an attempt to discover the causal facts or determinants of an event of interest. An example of an ex post facto investigation would be if there were an accident at a sports event and evidence was needed to understand why the accident happened and how such an accident could be repeated in future. The event of interest is the starting point for ex post facto research. The data to be recorded include measurements of objects, video recordings and statements from people related to the event of interest. Ex post facto research can also investigate the causes of some long-term present condition.

There are two types of ex post facto investigation: causal research and causal-comparative research. Causal research attempts to establish causal relationships between a present condition and antecedent conditions. In causal-comparative research, the causes of a condition of interest are studied by retrospectively looking at antecedent conditions for subjects possessing the condition and subjects not possessing the condition. This is similar to ex post facto types of quasi-experimental design, except that there are also non-experimental ways of undertaking ex post facto research.

DATA GATHERING AND ANALYSIS METHODS

The previous section was concerned with broad research types without specifying the nature of the data involved, how the data are collected or how the data are analysed. This section describes methods used to gather and analyse data that can be used within broad research processes. Figure 2.2 shows that data gathering and analysis methods can be qualitative, quantitative or, in the case of some questionnaires, can contain a combination of both qualitative and quantitative data.

Interviews

Interviews are used for many different purposes, but in research the aim of using an interview is to learn about the experience, motives, attitude, beliefs and opinions of the interviewee. They can be used within different types of research including historical research, survey research and case studies.

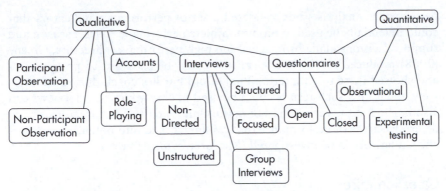

Figure 2.2 Data gathering and analysis methods

Interviews are typically done in person but it is also possible for the researcher to use a team of trained interviewers or for an interview to be conducted using telephone or Skype communication. Interviews are either audio or video recorded allowing a transcript of the interview to be produced and analysed. (The analysis of interview data is covered in Chapter 9.) Where the interview is video recorded, it is possible to analyse both verbal and non-verbal communication. The main advantages of interviews over question-naires are that they permit in-depth discussion of the topics of interest, probing, clarification and flexibility.

There are different types of interview that can be classified by whether an individual or group is being interviewed on each occasion as well as by the style of question and answer communication that is used. An interview could use a predefined set of questions with no scope for flexibility, com-pletely unstructured or semi-structured. Group interviews are useful as responses can be corroborated or challenged by other group members. The interviews are planned so that sub-topics are known, based on preliminary reviews of relevant literature. An interview guide is useful to ensure that all of the sub-topics of interest are discussed during the interview. There are also disadvantages to interviews in that they use a small set of participants and occasionally a single participant.

Interviews have been used in conjunction with performance analysis research. For example, Donnelly and O'Donoghue (2008) analysed the behaviour of coaches of three different levels of netball player and inter-viewed the most senior coach in their sample. The results showed that coaches of higher level players showed greater variability between sessions than coaches of lower level players. However, while the performance analysis methods could identify this difference, they could not explain the difference. Therefore, an interview was used to discuss the performance analysis results with the most senior coach to seek explanations for the differences found. Another example of the combined use of performance analysis and interview-ing was a dissertation by Greene (2008) on 400m hurdles performance. The performance analysis study validated a set of performance indicators that could potentially be used to monitor athlete performance over a season and support decisions made by coaches and high performance directors. Greene (2008) produced results in a form that could be provided to practitioners and showed these to an international 400m hurdles coach during an interview about how such results could be used in elite athletics. Interviews can also be used in the early stages of performance analysis research projects to determine areas of performance that can be worked into potential performance indicators to be investigated (McCorry *et al.*, 1996).

Questionnaires

Questionnaires are written or electronic forms to be completed by respondents for analysis by the researcher. Questionnaires allow a greater number of participants to be surveyed than interviews. However, questionnaires do not permit the level of probing or flexibility that interviews do.

There are many different types of questionnaire, including self-completion questionnaires, which could be administered in person or sent and returned by post. There are also questionnaires to be completed by the researcher while asking questions in the presence of the respondent or during a telephone call. Self-completion questionnaires have the advantage that the researcher does not need to be present when they are completed by respondents. This can help ensure accurate responses, especially if the questionnaires are anonymous. A disadvantage of self-completion questionnaires is that they require much greater development effort to ensure that they are complete, clear, well designed, well presented and provided with sufficient supporting instructions and incentives for the respondent. Questionnaires have four main sources of limitations that need to be recognised by those undertaking questionnaire surveys (Kirk-Smith, 1998). These limitations relate to:

1. theory building;
2. validity of self-report;
3. measurement; and
4. analysis.

Questionnaires involve question-answer communication that can be used to characterise or diagnose a problem and factors associated with it. However, in the absence of experimental intervention, there is no evidence to determine whether there is a causal link between any factors associated with the problem and the problem itself. This renders questionnaire evidence weak when it comes to developing theory relating to the problem of interest.

Questionnaires, like interviews, are a self-report and the link between questionnaire response and actual behaviour can be questioned. Respondents' (genuine) attitudes and beliefs may not predict their (genuine) actions. There are many factors that influence the accuracy of self-reports including image management and social desirability effect. Essentially, the respondent does not wish to look foolish when completing a questionnaire and will answer questions in a way that gives a positive impression. Questionnaires may relate to real-world behaviour contexts in different situations, but the questionnaire is completed by a participant away from that real-world context. Participants being asked what they might do if their bag were snatched by thieves on the street, might give what they consider to be an honest answer. However, in such a situation, the participant may not behave the way they believe they would. In answering the question, they may not have imagined the full scenario they are being asked about and may not realise how little they are aware about how they would actually behave. Another issue is that most behaviour is habitual, but answering a question requires more thought. Therefore, people may not know the causes of their own actions.

There are a number of measurement issues in the use of questionnaires that need to be considered. Investigator bias may be present in the selection of questions asked, the way in which questions are asked and the order in which they are asked. Respondents may perceive there to be an agenda in the research being done and the decisions that could be taken as a result of the study. This might influence how they respond to questions. The scales of measurement used for rating questions might not allow sufficient discrimination between different levels of the concept being rated. A further issue is that where questionnaires are used before and after some experimental period, it is possible that respondents may remember their pre-test answers when completing the post-test. This raises a question about the accuracy of the responses and if any pre-post-changes reported reflect actual pre-post-changes. Questions may not have been so relevant to the respondents during the pre-test, but after the experimental period, those questions have greater importance to the respondents and this differing context of questionnaire completion may influence the responses made.

The fourth area of limitation of questionnaires described by Kirk-Smith (1998) was the analysis of data. Researchers should define the purpose of the study and design the questionnaire in a way that analysis is done to answer the stated research question. Booth (cited in Kirk-Smith, 1998) used the term 'multivariate data grubbing' to describe analysis that some may be tempted to do, given the sophistication of data analysis packages and data mining routines available today.

Despite these limitations, which researchers should recognise, questionnaires are undeniably useful in many areas of research. Previous research in performance analysis has used questionnaires within part of the study (Blaze *et al.*, 2004, Hale, 2004, Jenkins *et al.*, 2007). Hale (2004) compared competitive and training matches in netball using time-motion analysis, interviews and the DM-CSAI-2 instrument. The purpose of the DM-CSAI-2 instrument within the study was to identify whether the intensity or direction of different types of anxiety differed between the two types of game. Blaze *et al.* (2007) surveyed 10 English FA Premier League clubs to describe the current use of performance analysis within elite soccer clubs. Part of the study done by Jenkins *et al.* (2007) provided motivational videos to a netball squad and then used a questionnaire to record players' beliefs and attitudes about the provision of motivational videos. Due to the limited number of players within the squad (n = 12), this questionnaire was composed entirely of open questions, allowing rich qualitative information about the usefulness of motivational videos to be analysed.

Questionnaires have been used as an alternative to observational techniques in monitoring physical activity. Montoye *et al.* (1996: 34–41) described exercise diaries and activity recall questionnaires, their validity, advantages and disadvantages. There is a trade-off between feasibility of data collection, reliability of measurement and number of participants that can be included in a study. It is neither feasible nor productive for an observer to watch a participant 24 hours per day to record those periods where moderate or high intensity activity is being performed. Observational techniques may be more reliable than data based on participants' recall of their physical activity, but there are still limitations to observational techniques. A greater number of participants can be included in a study of recreational activity if activity recall questionnaires rather than observational methods are used.

To date, there appears to have been a greater use of interviews than questionnaires within performance analysis of sport research. The reason for this could be that researchers using mixed methods like to use methods that are as complementary as possible, which favours the use of the most qualitative techniques.

Testing and measurement

Testing and measurement are data gathering activities that can be used within different types of research including experiments, observational studies, surveys, case studies and longitudinal research. Experimental studies will typically manipulate some independent variable to determine the effect on some hypothesised dependent variable. The independent variable could be whether or not some training programme is being undertaken over an experimental period. The dependent variables are typically tests of some ability or fitness characteristic. An example of such a study was a quasi-experimental study done by King and O'Donoghue (2003) into the effectiveness of a 13-week specific intermittent high intensity training session for under-14 county level Gaelic footballers. This study used a 20m sprint test, a vertical jump test, a multistage fitness test (Ramsbottom *et al.*, 1988) and a test of eight 40m agility runs (Baker *et al.*, 1993). These were performed by all participants before and after the 13-week experimental period. There are general tests of other attributes of fitness such as standing long jump, medicine ball throw, sit and reach tests, anthropometric tests (Hughes, 2008) as well as sports-specific tests for racket sports (Hughes, 2008) and field tests (Carling *et al.*, 2009: 115–26, 154–7). These tests have been developed, specified and validated within the sports science literature. The use of these fitness tests is not confined to pre- and post-testing in experiments, but fitness surveys can also incorporate these tests (NIFS, 1989). Tests of visual perception, anticipation and decision making have also been developed and applied within soccer research (Carling *et al.*, 2009: 43–69).

Observational research is a broad research type that can be undertaken using quantitative measurements or qualitative techniques. Where observational research involves quantitative measurement, measurements of aspects of real-world behaviour are made rather than performances within a controlled fitness test. The types of measure that can be made include counts of events, timings of behaviours and distances covered. Some of these measurements can be made with fully automated measurement techniques such as GPS equipment (Carling *et al.*, 2009: 91–3), mainly automated systems, such as ProZone3® or systems requiring extensive operator activity (Montoye *et al.*, 1996: 26-32). For recreational activity, pedometers and other movement detection devices can also be used (Montoye *et al.*, 1996: 72–90).

Field notes

Field notes are produced and analysed during qualitative research that involves field work. A researcher enters the field either covertly or overtly for the purpose of investigating the activity of a group of interest. Field notes have a disadvantage compared to interview transcripts because the interview transcripts contain data that come directly from the participants. Field notes may be written up after sessions in the field and are limited by the researcher's ability to recall the events that occurred. Spradley (1979: 74–6) recommended using four types of field notes:

1. Short notes made during the field episode – this is not always possible, especially if field work is being done covertly.
2. Expanded notes to be made as soon as possible after a session in the field.
3. A field work journal reporting problems occurring during field work.
4. A provisional running record of analysis and interpretation of field data.

The use of field notes is covered in greater detail in Chapter 9.

ASSUMPTIONS OF RESEARCH METHODS

Burrell and Morgan's framework

Sports science is a social science as it is concerned with investigating people: sportsmen, sportswomen, coaches, officials and spectators. Therefore, the way in which human activity is investigated is largely influenced by assumptions about the nature of human behaviour and conceptions of social reality. Research methods can be classified as objective and quantitative, or subjective and qualitative. The research methods are based on approaches that have underlying assumptions about human behaviour that researchers should be aware of. Burrell and Morgan (1979) described two approaches (or paradigms) that research methods follow: the normative paradigm and the interpretive paradigm. An excellent description of Burrell and Morgan's (1979) framework is provided in Cohen *et al.*'s (2007: 7–26) opening chapter and it is recommended that students undertaking any kind of research read it. Burrell and Morgan used a framework of the following four elements to compare and contrast the paradigms:

1. ontology;
2. epistemology;
3. human nature; and
4. methodology.

In this section, the first three of elements of Burrell's and Morgan's framework will be discussed, as methods are discussed throughout the remainder of the chapter.

Ontology is concerned with assumptions relating to human existence and experience. The normative paradigm is based on the assumption of realism and that objects exist in a real outer world that is independent of different people's views of the world and the objects in it. This means that this assumed real world and the objects within it can be investigated. Behaviour of people in this assumed real world can also be investigated assuming such behaviour can be portrayed in a manner that is independent of different people's interpretations of the behaviour. The interpretive paradigm is based on the assumption of nominalism, which assumes that people label things in their own mind but these labels are personal and not universal representations of objects that are external to the person's mind. Individuals are assumed to live in their own inner worlds with different views of issues and situations. The interpretive paradigm regards these individual world views as important enough to investigate in their own right. People experience life through their senses, reasoning and thoughts and assume the world is the way it is presented to them by their own senses. The human brain is a biological neural network that may learn to represent things in different ways in different people. None of us actually knows if the way images, sounds, feelings, tastes and smells are presented to our inner selves the same way as they might be presented to other people by their sensory perceptions. This leads to deeply philosophical questions about whether a single real world actually exists. The interpretive paradigm is an approach to research that is based on the assumption that individuals experience their own inner worlds.

Epistemology is concerned with how knowledge is represented and communicated. The normative paradigm is based on the assumption of positivism, which assumes that knowledge takes the form of objective tangible facts. Positivism assumes that knowledge consists of things that can be measured and classified. This is related to the ontological assumption of realism, as positivism assumes that a measurement procedure can be developed that can be applied by different people obtaining consistent results. The interpretive paradigm views knowledge as more complex and personal representations of experience, beliefs and attitudes. Knowledge in the form of pictures, words, sounds, emotions and feelings are difficult to quantify and different people can draw different conclusions from data. The author was going to write 'different people can draw different conclusions from the same data', but are a person's senses really presenting them with the same data as another person's senses? Such data are interpreted using personal opinion, which is related to the ontological assumption of nominalism. In sports performance, there are some situations where the assumption of positivism applies and other situations where antipositivism is more appropriate. Any weekend, when watching television coverage of football, a panel of analysts will watch the same video evidence of an incident and be aware of the same written regulations for the sport, but will still form different interpretations of the decision the referee should have made. In other situations such as tennis, there are fewer situations that are open to individual interpretation.

The human nature assumptions of Burrell and Morgan's (1979) framework are crucial to understanding the research methods that are based on the assumptions of the normative and interpretive paradigms. Human nature assumptions relate to how humans behave in different situations. The normative paradigm views human behaviour as deterministic, predictable and past-oriented. The assumption of determinism is that humans behave in a similar way in given situations and that we can reason about the average human being. This permits reductive techniques to be applied and generalisations to be made from samples used in research studies. The methodologies supported by the normative paradigm are referred to as 'nomothetic' and are empirical methods that represent samples by summary sample parameters such as means and standard deviations. The interpretive paradigm, on the other hand, views human behaviour as being more future-oriented, voluntary and individual. Voluntarism assumes that humans will take responsibility and exercise free will in how they behave in different situations. Therefore, the interpretive paradigm rejects the notion of an average human being and supports idiographic methodologies that recognise and explore individual differences. Some individuals may be naturally more deterministic than others in the way they behave, conforming to behavioural norms.

When one considers those who have careers in the armed services, determinism is essential. These individuals are part of an organisation with a task to perform where all must be trusted to obey orders so that the hierarchy of command will result in the task being accomplished. There are others who behave in a more individual manner, working in creative areas such as graphic design or the arts who will be allowed to behave in a more voluntary way to display flare and originality in their work. It is also common for the behaviour of some individuals to be deterministic in some situations and voluntaristic in other situations.

The normative paradigm

The normative paradigm is based on the assumption that there is a single real world that is independent of the view of individuals of that world. This not only applies to objects within the world but also to processes and rules that govern human behaviour. If behaviour is deterministic and predictable, then it follows that all humans behave in a similar way. Essentially, there is an assumption that there is an average human being that can be studied through an empirical approach. Human behaviour is abstracted to hard tangible facts that can be used to represent behaviour in a concise way, facilitating quantitative analysis. The quantitative methods used recognise variability about this average human being through measures of dispersion that are used along with measures of location when describing samples. The process of research involves the following steps:

1. Abstraction – representing human behaviour with a finger-print of categorical and numerical variables.
2. Sampling – gathering data using a representative sample of the population of interest.
3. Summarising – determining sample parameters to describe the average participant.
4. Analysis – using statistical procedures to determine relationships and associations between the variables used.
5. Generalisation – assuming that a large enough random sample is representative of the population of interest.

Figure 2.3 summarises the process by which the sample parameters (mean and standard deviation for example) are generalised to the whole population. This type of research does recognise the possibility of sampling error, where a sample derived from a population may show an opposite pattern to the population. Statistical procedures will be covered in Chapter 8; at this stage we will simply understand that the p value that is often reported in quantitative research is the probability of a Type I Error. A Type I Error is made when a difference found in a sample does not actually exist in the wider population. Whether or not a Type I Error has been made will be unknown to the researcher, but the chance of a Type I Error can be reported. As an example of sampling error, the author supervised three time-motion

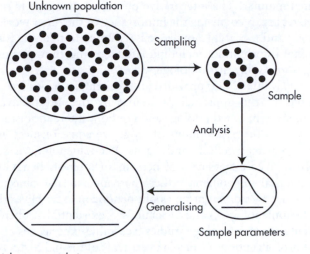

Unknown population

Sampling

Sample

Analysis

Generalising

Sample parameters

Unknown population parameters

Figure 2.3 Populations and samples in empirical research

analysis studies of men's Gaelic football in 1999, 2000 and 2001 that used identical methods and participant numbers. The first student found that midfielders spent a significantly greater percentage of match time perform-ing high intensity activity than defenders and forwards; the second student found no differences between these three groups; and the third student found that midfielders spent a significantly lower percentage of match time performing high intensity activity than defenders and forwards! At least two of these projects have a sampling error. However, all were done correctly because the two projects that reported differences also reported p values that acknowledged there was a small chance of a Type I Error, while the other student acknowledged there was a chance that he made a Type II Error. A Type II Error is where there is a difference for the population in reality that is not revealed when analysing the sample used.

The interpretive paradigm

There are many challenges to natural science approaches to social science. Human beings have an active consciousness and exercise choice. This is particularly relevant to performance analysis because sportsmen and sports-women are goal-oriented and ambitious. These individuals, particularly elite athletes, are unique with important individual characteristics and expe-riences that are lost from analysis where the research paradigm tries to sum-marise these different real individuals as an abstract average elite athlete. The interpretive paradigm rejects the image of human beings as passive determined organisms that respond to stimuli from their environment in a past-oriented manner. The interpretive paradigm also rejects the notion of an absolute reality, recognising the importance of the inner worlds that individuals create and modify. Knowledge is assumed to take the form of softer more personal information, including language and thoughts that are analysed using idiographic methodologies.

Cohen *et al.* (2007: 22–5) provided an excellent description of three different variations of the approach promoted by the interpretive paradigm: phenomenology, ethnomethodology and symbolic interactionism. Phenomenology is based on the assumptions of an active consciousness and an inner world where a person reflects and learns. Phenomenology is suited to the analysis of the lived experiences of persons of interest. Behaviour is determined by internal phenomena rather than an external objective physical world. There are two different types of phenomenology; Husserl's transcendental phenomenology and Schutz's existential phenomenology. Transcendental phenomenology studies face value experiences of phenomena rather than preconceptions or media portrayals of them. Existential phenomenology views experience as being an unbroken stream of consciousness rather than individual experiences that have little meaning on their own.

Ethnomethodology studies how people make sense of the social world that they experience and their social interactions. In particular, ethnomethodology studies group behaviour and cultures that have developed through the interactions of group members. Linguistic ethnomethodology studies the use of language, while situational ethnomethodology studies how people negotiate during the situations they experience.

Symbolic interactionism assumes that people's action towards objects and other people is determined by the meaning that they place on those objects. This is antipositivism whereby different people's views of objects is assumed rather than an absolute outer world that is independent of people's views. A person's inner world is assumed to be a dynamic concept that is continually constructed and modified as people experience different situations and weigh up the potential response of others as well as advantages and disadvantages of the behavioural choices they have. This is relevant in performance analysis of sport as it impacts on our understanding of how athletes and coaches think, learn and use information.

Critical theory

Cohen *et al.* (2007: 26–32) described a third paradigm called 'critical theory', which is relevant to the practical application of performance analysis. The normative and interpretive paradigms are viewed as incomplete as the political and ideological contexts of research are not addressed. Critical theory goes beyond describing behaviour in an accepted normal situation by realising changes to the situation. Cohen *et al.* (2007: 26–32) used terms such as 'emancipate the disempowered', which biomechanists and notational analysts would not typically use in their research and teaching.

However, when one considers the four stages of description, information, confrontation and reconstruction used in critical reflection (Smyth, 1989), critical theory is closer to the way performance analysts work in practice than the normative and interpretive paradigms.

The performance analyst working with a squad will analyse relevant aspects of squad performance, providing a description of the current situation. This would be the end point of descriptive research, but performance analysis in its coaching context has several more stages of a critical reflection and performance enhancement process. A thorough analysis of sports performance data identifies areas of the squad's performance that can improve. This challenges the status quo of the squad's situation within the sport, informs areas of 'illegitimacy' of the situation and threatens the interests of dominant squads. The confrontation stage is also relevant to performance analysis in practice as the squad is not so much asking if they can improve but how they can improve. The reconstruction stage is where decisions are made relating to squad preparation to specifically improve the performance of the squad. The effectiveness of this action is evaluated during analysis of training and subsequent competition.

The research methodology most underpinned by critical theory is action research, described earlier in this chapter. Critical theory has a transforming intention and seeks to ensure the catalytic validity of research, meaning that the research can be an agent of change in practice. This is exactly why performance analysis is used by coaches and athletes.

QUANTITATIVE RESEARCH PROCESSES

Figure 2.4 shows the 'V' shaped model of the quantitative research process being proposed in the current text. The idea for this model comes from the author's background in software engineering, where a 'V' shaped model is used by the European Space Agency to represent the stages in the development of a software system and the links between these stages (Robinson, 1992: 3). If the user requirements for a software product are not fully understood by the developer at the earliest stages of system development, a great deal of system development activity can take place before the developers even realise they are developing the wrong product. This is similar to research where mistakes made early in the research process may not be detected until after a great deal of data collection and analysis activity have been completed. The student or researcher undertaking a research project will have a broad idea of the research problem they wish to investigate, but typically keep this vague until after at least an exploratory review of previous research. The review of literature is a survey of previously published research in the area of interest that determines 'where' we (the scientific community) are with our knowledge of the topic. The review of literature does not only cover the state of the art in terms of theory but also in terms of relevant variables and methods. This places the researcher in a much better position to transform the broad and vague research problem into a more specific and detailed research proposal. The rationale for the study identifies a gap in our knowledge of the chosen area and a need for the study, justifying its importance. Therefore, the rationale explains 'why' the study is being done.

In quantitative research, formal hypotheses are used to state the specific research question in terms of precise variables and conditions. The hypotheses typically include a null hypothesis (that there will be no difference between samples or no relation between variables) as well as an alternative hypothesis (that there will be one or more differences between samples or relationships between variables). The hypotheses state exactly 'what' the research question is. The study to answer the research question is then designed. Methods are devised, describing 'how' data will be gathered and analysed to answer the research question. The fixed design is very common in quantitative research and permits data collection to be undertaken by trained personnel other than the researcher in some investigations. Once all of the data are gathered, they are analysed using statistical techniques that produce the results of the study. These results constitute 'what' has been found by the research study and whether it is the null or the alternative hypothesis that cannot be rejected. The discussion draws on literature from directly and indirectly related areas of scientific research to offer explanations for the findings. Theoretical perspectives and other research evidence are used to explain 'why' the study has produced the findings that have been observed. The conclusions do not merely summarise the results but also state 'where' our knowledge of the area is now that the study has been completed, providing directions for future research opportunities.

Figure 2.4 is very deliberate in using a 'V' shape to represent the quantitative research process rather than a straight line sequence of seven stages. The results are related to the stated hypotheses and the discussion is related to the rationale for undertaking the study as well as the literature review. Therefore, if the researcher does not thoroughly cover the literature, the entire study may be addressing a 'gap' in research that has already been covered by previously published research. This might not be detected by the researcher until the discussion is being done. This will mean that a great deal of project activity including research design, data gathering, data analysis and presentation of results will be done before the mistake is detected.

Similarly, when the hypotheses are formed, they must be precise and in terms of variables that are measurable and testable. The hypotheses should not only be precise, but should also dictate a results format that will comprehensively answer the research question while also allowing for a sufficiently concise presentation of results. The researcher would not wish to gather and analyse data, possibly taking hundreds of hours, only to find that

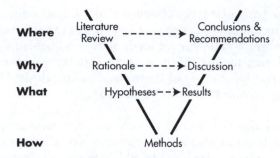

Figure 2.4 The 'V' shaped model of the quantitative research process

the results are trivial or that they are overly verbose. Therefore, planning in quantitative research is essential and is promoted by the use of a fixed research design. Although performance analysis does not involve organising testing, making arrangements with participants for testing and booking equipment to the same extent that some experimental studies do, good planning is still necessary for the delivery of a good dissertation.

At the earliest stages of conceiving a research idea, researchers should be able to visualise whether there are feasibility and logistical problems with the research. However, in the experience of this author, students still propose ideas that are not feasible or practical when this should have been immediately obvious to them. As soon as a student thinks of a research idea, they should ask the following questions:

- Is there public domain video footage of the performances that I need to analyse and am I able to obtain copies of this footage?
- Will public domain footage show the events that need to be studied?
- If public domain footage is unlikely to be obtained, is it possible to film matches myself? Will I get permission to film? Will I get access to a good filming position at the competition venue? Will I be able to book the video cameras I need to record data? Will I be able to travel to the matches? How much will the admission fee be at these matches?
- If I wish to film training sessions or coach behaviour, will enough players or coaches at the given level of the sport be willing to participate in the study? Does the training venue have a good filming position? Am I available to travel to the training venue and film sessions on the days the squad trains?
- If the study involves detailed analysis of technique, will I be allowed to use equipment in the biomechanics laboratory? Is such equipment only available to students taking level 3 or level 4 modules in biomechanics? Will I really be able to get relevant athletes to participate in the study?
- If filming cannot be done and there is no public domain video footage, can I use manual methods to record the necessary aspects of performance live?

- If there are particular types of match that I am intending to study (for example five-set tennis matches at Grand Slam tournaments or soccer matches where both teams are level, ahead and behind for at least 15 minutes during the match), how many matches are there that satisfy these criteria? How many of these matches are televised? Will this be sufficient to answer the research question?

Students should also give serious thought to what their proposed project will actually be like and whether the tasks they say will happen will indeed happen. If a dissertation idea is not feasible, effort in that research direction should cease as soon as possible and a more feasible project should be selected.

Cohen *et al.* (2007: 80–81) described orientating decisions that are made at the early stages of a project. These include determining the purpose, aims and scope of the research, identifying constraints and operationalising the aims of the research. Once a feasible research area is agreed, the project can be planned by identifying tasks to be done, estimating the time required to do them and associating tasks with deliverables to be produced. When devising methods in quantitative research, the opportunity should be taken to produce a detailed plan of the research project, which will act as a tool for project management and control. The various data gathering tasks, data analysis, reporting tasks, write up tasks, obtaining ethical approval and any other task that requires time and effort should be identified and listed. An initial timetable should be drawn up showing the weeks from when project activity commences to the deadline for submission of the dissertation. This plan should account for every project activity that requires time and effort including library work, data gathering, data analysis, writing activity and meetings.

QUALITATIVE RESEARCH PROCESSES

Some qualitative studies develop hypotheses during data gathering and analysis rather than testing hypotheses through a process of data gathering and analysis. However, this does not mean that such research projects are not planned and so there will be some purpose and scope to the research. Planning should address sampling, data gathering, analysis and establishing trustworthiness (Lincoln and Guba, 1985: 240–48). Cohen *et al.* (2007: 171) explained that while there is no single blueprint for every study that could be done, qualitative research should go though the following stages:

1. locating a field of study;
2. addressing ethical issues;
3. deciding on a sample;

 4. finding a role and entering the context;
 5. finding informants;
 6. developing and maintaining relations in the field;
 7. collecting fieldwork/data;
 8. collecting data outside the field;
 9. analysing data;
10. writing the research report.

Qualitative research has a more flexible design than quantitative research and emerging areas of enquiry can be investigated that may not have been anticipated at the beginning of the study. Chapter 9 will cover qualitative research in more detail.

MIXED METHODS APPROACHES

Using a single method may distort the theories that are developed as a result of the assumptions and limitations of that method. Mixed methods research combines the use of quantitative and qualitative research in a complementary way. There is a continuum of mixed methods approaches with varying contributions from qualitative and quantitative methods (Teddlie and Tashakkori, 2009: 94–6). The mixed methods approach might be dominated by quantitative research with a smaller qualitative study preceding or following the quantitative study. Alternatively, a small and efficient quantitative study using internet data may precede the extensive use of qualitative methods within the research project. The different arrangements lead to different research processes that include aspects of planning, data collection, analysis and interpretation that have already been discussed in relation to quantitative and qualitative research. It is very important that the project is not disjointed and there should be a natural link between the quantitative and qualitative elements of the study. Fully integrated mixed methods involve interdependent quantitative and quantitative elements where each type of data undergoes transformations to allow direct verification of the other type of data (Teddlie and Tashakkori, 2009: 280–81).

Some mixed methods performance analysis research has been done using notational analysis techniques alongside accounts (Brown, 2005), interviews (Brown, 2005, Donnelly and O'Donoghue, 2008, Greene, 2008) and qualitative observation (McLaughlin and O'Donoghue, 2004, Paisey and O'Donoghue, 2008). When instruments such as the Arizona State University Observation Instrument (Lacy and Darst, 1984) are used to analyse coach behaviour, quantitative data are recorded about the coach's behaviour in isolation from the interaction of players. The findings of these quantitative studies can serve as a starting point for qualitative analysis to explain the coaching behaviour used. This could be done using interview techniques (Donnelly and O'Donoghue, 2008) or qualitative observation of coaching sessions (Paisey and O'Donoghue, 2008). Similarly, McLaughlin and O'Donoghue (2004) used computerised time-motion analysis and qualitative analysis of video recorded behaviour in a complementary way to investigate the activity of primary school children in the playground. The quantitative time-motion analysis revealed that boys spent a significantly greater percentage of time performing high intensity activity than girls. The main limitation of the time-motion analysis was that a child's activity was analysed in isolation from the other children that he or she was interacting with. The qualitative analysis of the same video recordings allowed the researchers to explain gender differences in the activity performed by looking at how the child interacted with other children in the playground.

SUMMARY

Quantitative and qualitative methods can be used to gather and analyse data. Quantitative methods have underlying assumptions of social reality based on the normative paradigm. Qualitative methods, on the other hand, are based on the assumption of the interpretive paradigm. These assumptions have implications for how knowledge is represented and communicated, and each approach has advantages and disadvantages over the other. Quantitative and qualitative methods can be used within different broad types of research, which have been described in this chapter. There is no one single process that is used within all performance analysis research projects, although most quantitative studies follow the 'V' shaped model presented in this chapter.

CHAPTER 3

SELECTING A RESEARCH QUESTION

INTRODUCTION

When selecting a research question in performance analysis, the student will go through a process of selecting candidate areas of interest, before choosing one of these and turning it from an area of general interest to a specific research question. The student will investigate previous literature in the area to approach the research topic being well informed about theoretical and methodological aspects. This is the first stage of the research process and a considerable amount of project effort depends on directions set at this point. If the student selects a poor research question, this might not be detected until they are drawing conclusions and recommendations, assessing where our knowledge of the subject area is as a result of the study. This chapter is made up of two broad parts: the selection of a research area of interest, and reviewing previous research in the area.

CRITERIA FOR SELECTING A RESEARCH TOPIC

The criteria for selecting research questions in performance analysis are similar to the criteria for selecting research questions in other disciplines of sports science. The criteria for selecting research questions in sports science are (Gratton and Jones, 2004: 39–41, Thomas *et al.*, 2005: 26–7):

- the personal interests of the student;
- the career aspirations of the student;
- the research interests of staff in the student's university department;
- topical issues in sport;

- scientific literature on the area;
- practical application of research in the area;
- importance of the area;
- the strengths and weaknesses of the student.

Programme regulations

Students considering performance analysis research questions must first consider whether they can do performance analysis research or if they would be better undertaking research in another discipline of sports science. The student's programme of study may be sports management or sports development where staff and external examiners would expect the dissertation to be within the scope of the programme. Students should seek advice from programme directors if there is a doubt as to whether performance analysis is a relevant area to the degree qualification they are working towards. This author has not come across a situation where a student has undertaken a dissertation, done a good academic study and written it up well only to be advised that the dissertation cannot contribute towards their degree. However, there are programme regulations and this is a situation that students and programme administrators would wish to avoid. Where a student is undertaking a sports management programme (for example), it may be possible to use performance analysis within a sports management context. An example of this was a dissertation into 400m hurdles performance that investigated how performance indicators could be used by high performance directors to monitor athlete progress and make funding decisions (Greene, 2008).

Where performance analysis is an area within the scope of the student's programme of study, there may still be barriers preventing the student from undertaking a performance analysis research project. For example, some programmes may have regulations requiring students to take a level 3 depth option module in performance analysis in order to be able to undertake a performance analysis dissertation. Others may require the successful completion of a level 2 module in performance analysis. There are some universities that may not have entire modules devoted to performance analysis but which may include performance analysis within other modules such as coaching science modules. Whether or not there are regulations restricting the student's choice of dissertation is something the student needs to be aware of when selecting a research question.

The personal interests of the student

A research project may be a 40 UK credit point module at undergraduate level and a 60 UK credit point module in taught postgraduate programmes. The national standards for university modules indicate that there should be 10 effort hours of academic work at the module level for each credit point of the module. Therefore, an undergraduate research project will involve 400 effort hours of level 3 academic work by the student, while a Master's research project will involve 600 effort hours of level 4 academic work. With many universities running undergraduate dissertations over a 30-week period (September to March inclusive), the student will need to devote an average of 13 hours per week to their project in addition to other academic work they will be doing as part of their programme of study. Furthermore, the research project is an individual research project being undertaken by the student under the supervision of an academic member of staff. Such a sustained independent effort by the student requires a high level of motivation and commitment. If a student is undertaking a project in an area where they do not have a great personal interest, they may find it difficult to maintain progress. Where students make little or no progress in their research project for the first four months of the academic year, it is often because they find their project less interesting than reality television, social networking internet sites, computer games and socialising.

Some students may have decided to do a project that was an idea of the supervisor's rather than their own. Others may decide to do a dissertation in an area because they want to be supervised by a particular member of academic staff. There are many reasons why some students may elect to do a dissertation that they do not have a personal interest in. Students are advised to find an area that they are genuinely interested in, that will develop into a research question where they have an interest in the research problem and the eventual answer that their research project will provide. Some students may genuinely not know what areas they will be interested in at the time they submit a research proposal. One approach to starting the process of identifying an area of interest is to identify a sport that they are particularly interested in and then an academic discipline (which may not be performance analysis) that they enjoy working in. They should ask how the academic discipline can help their and the wider sports community's understanding of the sport. In considering sub-topics of the academic discipline, the student can gradually move towards more specific problems, asking what it would be really useful to know about the sport of interest that is not already known. Developing such an interest in a research topic will hopefully motivate the student to maintain the required effort over the 30 weeks of the project, leading to a successful outcome.

The career aspirations of the student

Many students on undergraduate university degree programmes have developed career aspirations before they reach the final year of their programme of study. These career aspirations influence module choices, work experience that is done, as well as the topic for their level 3 research project. Some career paths may involve undertaking specialist postgraduate programmes such as a PGCE (Postgraduate Certificate in Education) where there may be a high demand for places. A student wishing to embark on a PGCE pro-gramme, after graduating from their undergraduate degree programme, often needs to apply for the PGCE programme during the final year of their undergraduate degree. The selection process for such a programme is often very competitive and students are aware from an early stage of their under-graduate degree that they need to develop their curriculum vitae (résumé) to improve their chances of being given a conditional offer of a place on a PGCE programme. The student will undertake relevant modules in physical education, work towards and achieve recognised coaching qualifications and do work experience supporting physical education teachers in schools. The level 3 dissertation is also a way of furthering the student's interest and experience in their chosen vocational area. It is something they are able to include on their curriculum vitae and discuss during the interview stage of the PGCE application process. The student may have other genuine interests in theoretical aspects of sport and exercise science or practical coaching problems in the sport that they do. Sometimes, the student is able to combine their career interest and academic interests within a research problem to be investigated within the level 3 dissertation. Other students may not be able to combine these areas without the project appearing disjointed.

Where it is not possible to combine the student's career interest with their academic interests, the student has to consider the advantages and disad-vantages of undertaking research projects in their different areas of interest. The decision as to which area to follow for the level 3 dissertation depends on the short- and long-term benefits of studying in these areas. The short-term benefits include the grade the student would expect to be awarded based on their perceptions of their own academic strengths and weaknesses and the resources available to support their study in a given area. The long-term benefits include career opportunities that can be opened up through experiencing a given area of study.

A good example of performance analysis research applied to a vocational area is the study of school teacher behaviour. There are standard instru-ments of evaluating teaching style that have been used in educational research (Goldberger, 1992, Goudas *et al.*, 1995, Griffey, 1983, McBride, 1992). However, the use of instruments traditionally used for analysing coach behaviour allow physical education teacher behaviour to be com-pared to coach behaviour and discussed in relation to the behaviour of coaches. Therefore, Paisey and O'Donoghue (2008) used the Arizona State University Observation Instrument (ASUOI) (Lacy and Darst, 1984) to study the behaviour of physical education teachers during lessons in second-ary level schools. This research allowed interests in physical education teaching and coaching style to be pursued, while at the same time develop-ing practical skills in the use of computerised video analysis packages, reli-ability evaluation and data analysis, developing awareness of ethical issues in the study of children and developing the organisational skills involved in project planning, management and control.

The research interests of staff in the student's university department

Members of staff in university departments will have research interests and very often these are displayed on departmental web pages. Furthermore, research-active staff can deliver sessions to level 2 students on research topics that assist them in making decisions about what they will do for their level 3 dissertations. In some universities there are specialists in performance analysis who not only teach in the area but undertake personal research as well. In other universities, performance analysis dissertations may be supervised by academic staff whose primary interests are in other areas. Therefore, it is not always possible to develop a research idea in performance analysis from the research interests of academic staff. However, staff will have done sports science degrees themselves at some point in their past and have a sufficient knowledge to supervise outside their own discipline. The choice of supervisor is often outside the student's control as there may be a higher demand for some staff than their capacity to supervise. Students doing time-motion analysis dissertations are best being supervised by sports physiology staff where a specialist performance analyst is not available. Where a tactical analysis research project is being done, a member of staff who has an interest in the particular sport, or who coaches or has other interests in related sports could supervise. Alternatively, staff with interests in coaching science can supervise these projects based on their knowledge of information used in feedback to athletes. Students undertaking analysis of technique projects can be supervised by members of staff who teach biomechanics.

It is not always possible to develop a research idea based on the research of staff in the student's department and this is just one of the ways of identifying a research question. Indeed, it may be unwise to identify a research question that closely resembles staff research interests if the area is not an area of genuine interest to the student. The following list contains some of the common mistakes students make when basing a research question on staff research interests:

- Choosing an area to try to get a supervisor who is an expert at statistics – this is unwise as the research project is a sustained piece of work to be undertaken by the student and the supervisor will not do their statistical analysis for them.
- Choosing an area where the university has a very good track record in research performance – this is unwise if the student lacks ability in the area as the staff involved in the area will set high standards and expect the student to view the area in the same way as they do – The Holy Grail.
- Choosing an area to try to get a particular member of staff as a supervisor – it is better to choose a project for the right reasons. If the member

of staff leaves the university in the summer between the second and third year of the student's programme of study, will the student still wish to do the chosen research project?

Topical issues in sport

Interesting research areas can be initiated by non-academic sources such as media hype or personal experience. A good example of a topic arising from media coverage was the research undertaken by Rowlinson to compare player performances in the 2002 World Cup and the UEFA Champions League (Rowlinson and O'Donoghue, 2009). This started with a statement by the Manchester United manager, Sir Alex Ferguson, in an interview in April 2003, that the World Cup was not as high a standard of competition as the UEFA Champions League: 'I think the European Champions Cup is now bigger than the World Cup. All the best players are in Europe now' (http://www.abc.net.au/soccer/items/s827276.htm). This was something that captured Rowlinson's imagination and he started to ask himself if Sir Alex Ferguson was correct and how he could go about showing whether or not this statement could be justified. However, before he could embark on the study he wanted to be sure that this was an acceptable topic for a dissertation on a Master's programme in performance analysis.

In academic research, the statement of Sir Alex Ferguson alone would not provide a sufficient rationale for the investigation. However, there is descriptive research in the performance analysis literature where the quality of play is compared between different levels of the game using indicators of technical effectiveness. This research is open to criticism on methodological and theoretical grounds and the approaches used have limitations in relation to the study that was being proposed by Rowlinson. The role of performance analysis itself in answering such a research question can also be discussed in relation to alternative methods such as survey techniques. Those players competing in both tournaments as well as those players competing in one or other tournament can be identified and spectators could be surveyed to rate the quality of the players. This is a valid means of gauging supporter perceptions of the quality of players competing in the finals, semi-finals and other stages of the two tournaments.

This research problem, like many others, can be investigated using a variety of different research methods. Performance analysis is one of the disciplines of interest to this particular research area and there are many different ways to undertake a performance analysis investigation to compare the two tournaments. Each way of doing the study will have strengths and limitations, which should be recognised. These strengths and limitations can be discussed using literature relating to different approaches used in performance analysis for data gathering, processing and analysis. The project could compare one set of players competing in World Cup competition with another set of players competing in UEFA Champions League competition.

Alternatively, the same set of players could be compared when they were competing in the World Cup and when they were competing in the UEFA Champions League. This choice of independent or related samples triggers methodological debate from the outset of the project. Each player could be analysed using one performance from each tournament or, to avoid individual match effects having an impact on the study, each player could be considered using a profile based on multiple performances in each tournament. This draws in literature relating to the debate about whether it is better to use more players or more matches per player in such an investigation. There are no standard dependent performance indicators for technical effectiveness in soccer with different performance indicators being proposed by different researchers (Gerisch and Reichelt, 1993, Hughes and Probert, 2006, Olsen and Larsen, 1997). It is essential for the research topic identified through media hype to be supported by an academic rationale. In this particular case, the interest in the statement made by Sir Alex Ferguson generated an interest in many aspects of scientific research and underlying theory that could be used in such an investigation.

Scientific literature on the area

Reading academic literature in performance analysis can give the student ideas for potential research projects. During level 1 and level 2 modules, areas of different disciplines of sports science are covered. The content often includes research material, classic and contemporary theory, methodological developments and practical application of research findings. Sports science is a relatively new science and performance analysis of sport is an emerging discipline within sports science. This allows relatively original research problems to be identified. There is a range of originality of research that can be done, and it is very rare that any research is completely original in its contribution to our knowledge of sport. In performance analysis, what we mean by 'relatively original' is that analysis techniques applied to other sports in the past are applied to new sports for the first time. Alternatively, research done in the past can be replicated but using different levels or different age groups within the same sport, or analysing sport in the student's own geographical region or after a rule change has occurred in the sport.

An example of this is the use of time-motion analysis to assess the physical demands of different sports. Beck (Beck and O'Donoghue, 2004), for example, had read how computerised time-motion analysis had been used to determine the distribution of match time among different locomotive activities in sports such as netball (Loughran and O'Donoghue, 1999), basketball (Devlin et al., 2004), soccer (O'Donoghue, 1998), Gaelic Games (O'Donoghue et al., 2004a), field hockey (Huey et al., 2001) and rugby union (Knox and O'Donoghue, 1999). Beck's particular sport of interest was rugby league, and he wished to determine similar information for this sport to that which had been determined for other sports. The student had an interest in determining what was involved in playing a match, the intermittent nature of high intensity activity that was done and providing information that could be of practical use to those developing conditioning training programmes for rugby league players. There was nothing original about time-motion analysis or the specific use of the computerised time-motion analysis technique that Beck wished to use. However, his idea, which came from his survey of literature, combined with his interest in rugby league training led to the identification of a research project he could do.

When considering originality of research, the student should understand that there are disadvantages to undertaking research that is completely original in approach. The main disadvantage is that the number of directly relevant references they will be able to use in their eventual dissertation could be limited, requiring the use of indirectly related material. Where a dissertation idea takes a standard method, as Beck did, and applies it to a new sport, there are also disadvantages. Projects can be seen as becoming repeatable with very similar methods chapters, the same results format and the discussion of the physiology of intermittent high intensity activity can be very similar between independently written dissertations. Students undertaking such projects need to be aware of these potential problems and think of ways of adding greater originality to the project. One way of doing this is to introduce independent factors into the project. In the example of time-motion analysis of team games, there are factors that can be hypothesised to influence work rate and the distribution of match time among locomotive movement types. These include venue effects (Devlin et al., 2004), gender (O'Donoghue et al., 2004b, Huey et al., 2001), level of the sport (O'Donoghue et al., 2001, Knox and O'Donoghue, 1999), positional role (McErlean et al., 2000, Robinson et al., 1996) and score line effect (O'Donoghue and Tenga, 2001, Shaw and O'Donoghue, 2004). It is even possible to compare the work rate of different sports, particularly where it is common for players to compete in two different sports. Where one or more of these factors is included in the purpose of the study, it allows greater scope for originality in the eventual dissertation write up. Students should be careful not to include too many of these factors, however, as each factor can double (if the factor is measured at two levels, triple if measured at three levels, etc) the volume of data required for the statistical analysis.

Practical application of research in the area

The discipline of performance analysis typically involves analysing actual sports performance or at least analysing skills of the given sport in a laboratory-based study. Performance analysis is a highly sought after area of sports science support by many sports governing bodies. Therefore, many types of performance analysis investigations result in findings that are useful to those making decisions about the preparation of teams and individual athletes. The previous section used time-motion analysis as an example of a research idea that could come from reading previous research in performance analysis. Time-motion analysis is also an area with practical application in the conditioning of players. The findings give an understanding of the nature of competition, the duration of periods of activity, the duration of recovery periods and the type of high intensity activity that is performed by players. This knowledge alone would not be sufficient to recommend a training programme as this would require experimental evidence that undertaking the training programme improves performance. However, the knowledge provided by time-motion analysis does have practical application in the initial design of experimental training programmes.

There are many other areas of performance analysis that have practical application including tactical analysis, evaluation of technique, assessment of injury risk of activity, evaluation of referee performance and analysis of coach behaviour. A benefit of choosing an area that has practical application is that the potential use of the findings is a fruitful area for discussion when the results of the study have been determined.

As already covered in Chapter 1, the effectiveness of performance analysis support is itself a research topic. There are still many who are sceptical about the value of providing performance analysis support in coaching. The contribution to the body of evidence relating to the effectiveness of performance analysis has a clear practical application in that the evidence from the student's project and other research can be considered by high performance directors and coaches when making decisions about whether or not performance analysis support is to be provided. Two undergraduate dissertations together investigated the effectiveness of performance analysis support to a university netball team (Jenkins et al., 2007). Video recordings were made of the team's performances and used by the two students to provide two different types of feedback: instructional feedback (Morgan, 2006) and motivational feedback (Jenkins, 2006). Jenkins used qualitative techniques to evaluate player attitudes and responses to the provision of motivational videos before matches, while Morgan used a combination of quantitative (team performance) and qualitative analysis (of how the statistical and video feedback provided was used to make decisions about preparation for future matches).

For any performance analysis system or for the results of any performance analysis research to be applied in practice, coaches and other users need to be able to interpret the results. Therefore, recent research in performance analysis has developed a means of determining percentiles for different types of performances. Players' and teams' performances are influenced by the quality of the opposition they face, and O'Donoghue (2006b) proposed using different sets of percentiles for different levels of opposition. The understanding of what is a high value, an average value and a low value for different performance indicators in different types of match is essential for those practitioners looking for the areas of performance requiring the greatest attention. So far, these types of norms have been developed for

British National Superleague netball (O'Donoghue *et al.*, 2008), international netball performance (Blucher and O'Donoghue, 2007) and women's singles tennis at Grand Slam tournaments (O'Donoghue, 2008c).

Importance of the area

In performance analysis, the importance of a research area is related to the practical application of research findings generated by research in that area. The importance of the research area is not merely one of a number of different ways of identifying a research question. All students planning to undertake research projects in any area should seek to identify a worthwhile research question rather than a 'who cares' one. An example of an important research question is describing performance in a sport since some rule change (Williams, 2008) or since the introduction of new equipment (Brown and O'Donoghue, 2008c). While the direction of any difference may seem obvious to some, the precise extent of the change in performance is unknown without scientific evidence.

A good example of such research is the effect of the introduction of new ball types on tennis performance. The strategy of male and female tennis players at Grand Slam tournaments had been determined by O'Donoghue and Ingram (2001). However, the data used in this study were collected from Grand Slam tournaments between 1997 and 1999. Therefore, developments in player preparation and coaching techniques may have rendered O'Donoghue and Ingram's (2001) study out of date, even without the introduction of new ball types. In 2002, the ITF (International Tennis Federation) formally introduced the Type 1 and Type 3 tennis balls in addition to the Type 2 balls previously used. A formal classification of court surface pace was also introduced at this time to decide on which type of ball to use on each surface. The Type 1 tennis ball is harder and more resistant to compression upon impact than the standard Type 2 ball, leading to a reduced contact time with the surface and a faster game. This ball has been designed to be used on the slower court surfaces such as clay. The Type 3 ball is 6–8 per cent bigger than the standard Type 2 ball and generates greater air resistance, resulting in greater deceleration as it flies through the air. This ball is used on fast surfaces such a grass. The introduction of the new ball types was an attempt by the ITF to reduce the variation in the game between different surfaces. While it may seem obvious that the introduction of the new ball types may have reduced variation in tennis between the different Grand Slam tournaments, the precise effect of using the new types is unknown without scientific evidence. Therefore, Brown and O'Donoghue (2008c) investigated Grand Slam singles tennis comparing gender and surface effects using matches from 2007 Grand Slam tournaments. This was an MSc project that provided useful findings that would assist those preparing players for competition at Grand Slam tournaments played on different surfaces. Since this project was done, the surface used at the Australian Open has changed so that the same surface is used at both the US Open and the Australian Open. As further changes are made to equipment and regulations in different sports, it is important that research is done to provide an understanding of performance in those sports under the new circumstances.

The strengths and weaknesses of the student

As students undertake assessed work and examination in different areas, they will experience different levels of difficulty between different modules. Furthermore, feedback on assessed work will identify areas where they are strong and areas where they need to improve.

When considering their strengths and weaknesses, students can consider weaknesses in two ways. First, a weakness can be seen as an area where the student will lose marks in a research project. Where students are uncertain about future career directions, they often choose to undertake research projects in areas where they believe they can achieve good marks. A second way in which a student can consider weaknesses are areas where they need to improve. The whole point of the provision of feedback about assessed work is to guide the student in helping them improve in those areas. Students should take such feedback on board, improving where necessary to widen the number of potential areas in which they can work competently. There may be some areas where the student is unable to improve known weaknesses to a satisfactory level due to modules taken during their previous years of study. For example, the use of sophisticated systems for biomechanical analysis of technique may prove to be a barrier preventing data collection from even starting if the student does not have the ability to use such systems. Where a student has a genuine interest in undertaking a dissertation using such equipment, the student should make an honest decision about the research area based on their ability to undertake such research.

There are some abilities that are essential to undertaking a research project that all students need to develop, especially if these are areas requiring improvement. The ability to write scientifically is essential to producing the final dissertation report and students should access examples of good dissertations and read published research studies to improve their understanding of how methods are described, results presented and findings are explained in discussion sections. Students who are weak at scientific writing often make common mistakes such as displaying raw SPSS results or even sections of spreadsheets used during data analysis within their results sections. This type of presentation of results is never done in published research papers and students who have read published research papers in the library will realise that the outputs of data analysis packages need to be further processed, identifying the key results, and these have to be presented in the most reader-friendly way to communicate the findings of the study.

Another ability required by the student to undertake a dissertation is the ability to plan the project over a period of seven months, for example, identifying the tasks to be done, the associated deliverables and estimate how much time is required for each. A good project plan will recognise busy exam periods and assessment hand-in points where progress on the research project will not be as great as at other times. The research plan is a timetable of activity and associated deliverables that can be used by the student and the supervisor to manage and control the project. General time-management ability is also important and if a student has been weak in this in the past, they should use the research project as an opportunity to improve in this area, developing skills that are transferable to other areas of their life.

Some areas where students perceive themselves to be weak are where the student may actually have genuine aptitude that has never been realised. It has always amazed this author how students convince themselves that they cannot do statistical analysis before research methods modules even commence. Others do not see the relevance of learning about statistical analysis until they commence their level 3 research projects. The volume of knowledge that students need to develop during research methods modules is typically far less than that required in physiology, psychology, biomechanics, sociology of sport and other disciplines. There is nothing unusual about leaving a lecture or practical session in a module having failed to fully understand the content. Modules specify the expected number of effort hours required to achieve the learning outcomes of the module and these effort hours include necessary private study where material is read and exercises are undertaken to improve practical and intellectual abilities. Students at university are undertaking academic programmes such as BSc Honours degrees. These programmes are more than vocational and practical courses: they must include elements of science and hence must cover scientific research methods.

There are other areas where students may perceive their ability to be better than it actually is. For example, some students choose to do interview studies because they believe that qualitative research is easier than quantitative research. Qualitative research, in the view of this author, is more difficult to do than quantitative research. When using inferential statistics, a computerised package will determine whether a hypothesis must be rejected or not. In qualitative research, the researcher must analyse rich qualitative data, interpreting meanings from the data gathered based on knowledge of relevant theoretical concepts. Establishing the trustworthiness of qualitative data, being able to analyse responses during interviews to determine the next question and make decisions about areas to probe into, is not a straightforward process. Students should guard against having a naïve view of qualitative research and their ability to undertake good qualitative research in comparison to their ability to use statistical analysis procedures.

Feasibility

A key criterion in selecting a research question is that the research proposed must be workable (Thomas *et al.*, 2005: 28). This is a message that research methods lecturers and lecturers within other discipline areas try very hard to get across to students when discussing potential research projects. However, despite the very best efforts of these staff, every year in every sports studies university department, students propose research projects that are impossible to undertake. There are some research questions that are very interesting and important, but the projects proposed have major logistical and practical problems that cannot be overcome. A project that is not feasible needs to be abandoned or altered so that it becomes feasible before the student spends too much time and effort on it. Some examples of unworkable projects in performance analysis are as follows:

- Time-motion analysis of rugby player work-rate at national league level compared to the rugby world cup. The problem here is that the student needs player-cam type video rather than the on-the-ball type footage shown on television. Travelling to the matches in person to verbally code player movement sequences using a dictation machine is out of the question on cost grounds.
- Analysis of All-Ireland Championship Gaelic football. This tournament runs over the summer months and the student's dissertation typically takes place between October and March. If the student failed to collect data over the previous summer and there is no access to video recordings of the matches, then it will not be possible to collect data for the project.
- Analysing tennis player strategy in the point after successful and unsuccessful challenges are made using the Hawkeye system at Grand Slam tournaments. With four conditions of interest and a maximum of three unsuccessful challenges per set, there will be a very limited volume of data to be analysed in the proposed project. Furthermore, the student may have to watch sets that last over an hour to gather data from very few challenges.
- Biomechanical analysis of aerial challenges performed during competition would provide very useful results. However, any kind of quantitative analysis is impossible as such analysis would require a controlled laboratory set-up for the data collection.
- Analysis of young players' performance profiles to determine if any aspects of youth performance are associated with successful participation at senior level. The student's programme of study will have completed many years before the participants have become senior athletes.
- The effectiveness of a training programme for improving FA Premier League team performance in penalty shootouts. The student will not be able to perform any kind of invasive study using players at this level. They will be impossible to access.

Occasionally, a supervisor has to step in and put a stop to what should obviously be recognised as an unworkable project. While the research project is supposed to be a sustained piece of independent research undertaken by the student, this is one of the situations where supervisor intervention is essential.

LITERATURE REVIEW

Literature review process

Before commencing a research project in earnest, it is necessary to select an interesting, important and feasible research problem. The student should then review the available academic literature and possibly other literature relating to the research problem. The purposes of undertaking a review of literature are to identify a research problem, develop hypotheses and develop methods (Thomas *et al.* 2005: 29–31). The research problem may have been identified through other means, for example personal interest, media interest or career aspirations. Gratton and Jones (2004: 51–2) described some additional purposes of undertaking a literature review, which include demonstrating knowledge of the research area, understanding of relevant theoretical concepts and determining the extent of previous research. Gratton and Jones also listed the identification of variables, which would be included in the process of formulating hypotheses identified by Thomas *et al.* (2005: 55–7), and that the previous findings surveyed by the student could be compared with the student's eventual findings. The purposes of undertaking a literature review in sports science research in general also apply to the discipline of performance analysis. While performance analysis is an emerging discipline and theoretical concepts of sports performance may not have been fully developed, there are still relevant theoretical concepts from other disciplines that performance analysis overlaps with, which should be covered by the student.

Thomas *et al.* (2005: 32–49) described the literature review as a process consisting of the following six stages that lead to the eventual end product of a written review of literature:

1. writing the problem statement;
2. consulting secondary sources;
3. determining descriptors;
4. selecting preliminary sources;
5. reading and recording the literature;
6. writing the literature review.

Getting organised

Students should be properly prepared to review previous literature in a manner that is reasonably systematic, while allowing the necessary flexibility to follow areas of previous research that they discover to be relevant which they had not anticipated. The student needs to record details of articles that are read so that they can be used within the eventual review of literature chapter. Having a notebook dedicated to the literature search is a good idea and the students should bring this with them every time they go to the library. This avoids situations where the student finds relevant material but has to make a repeat visit to the library to record it or, worse still, forgets to record the paper or even where they found it. When the student has read a relevant paper, the title, author(s), journal/book details, publisher, volume, issue, page numbers and year/month should be recorded. A summary of the research described in the paper should be written down, so that the student will understand what was done and what was found well enough to be able to explain this to others in their own words. The main evidence provided by a paper is the results, and critical consideration should be made of the methods used and any limitations of these. The introduction and discussion sections of papers should also be read as these draw the student's attention to relevant theory and other evidence that may explain the results found. This information should be summarised by the student within their notebook and a list of key references used by the paper should also be recorded. This helps make the search for further research papers much more efficient.

Research papers and other material that are read by the student will vary in their relevance to the student's own study and so the volume and detail of what is recorded will also vary from paper to paper. There are some papers that will be so important that the student should photocopy them and keep within a dedicated folder of background material for their project. Other papers will be ones that the student will mention in their literature review but which may not be described as critically as others. As the student is conducting their literature search, they should periodically update an electronic reference list they are maintaining. Writing references is much harder than writing natural text and it is not a task the student should be starting during the final week of the project, so the earlier this is done the better.

Consulting literature

There is a wealth of sports science research in general and performance analysis research in particular that has been published in academic journals. On entering the journal area of the library, it is apparent that there are thousands of papers within these journals. The student would, therefore, be unwise to sequentially search through contents pages of different journals for articles that may be relevant to their own research. The student should instead use a systematic strategy for identifying relevant literature. The student should initially consult secondary sources such as research review articles and chapters of textbooks. Research reviews are not original research investigations but papers that survey the research done in a particular area. There are journals such as *Sports Medicine* and *Exercise and Sports Science Reviews* that only publish research reviews. Other journals that usually publish original investigations may occasionally publish review articles.

An excellent example of a review paper in performance analysis of sport is a review of automated and semi-automated motion tracking systems used in elite soccer (Carling *et al.*, 2008). This paper surveys the various commercial systems that are available as well as some other systems developed by academic institutions and used within time-motion analysis studies. The description of the underlying techniques used by these systems explores their strength and limitations with respect to the critical measurement issues of validity, objectivity and reliability. The paper also surveys motion analysis research that has been done using these techniques and the efficacy of their use as indirect measures of physical demands and fatigue in soccer. The authors also considered issues such as standardisation of variables used in such research and provided guidelines for the evaluation of the quality of measurements made by such systems. The paper refers to 92 sources, including previously published academic papers, papers that were 'in press' at the time of writing and websites containing information on commercial products. There were 62 of these sources that were dated in the five years between 2004 and 2008 when the review was published, making this paper an excellent secondary source for students studying movement in sport or the techniques involved in automatic player movement.

Textbooks are another type of secondary source and their advantage is that they are written by eminent people in the given field who are in a position to discuss the state of the art in different areas covered within the chapters. A good example of a book chapter that is a secondary source that could be used by students in the early stages of their literature reviews is the book chapter on 'Rule changes in sport and the role of notation' by Williams (2008), which discusses the role of rules in sport, the various motivations for making rule changes as well as research into the effectiveness of rule changes that have been made. This chapter is not only a useful secondary source for students doing research projects on rule changes in sport, but is also a very good example of how to structure a review of previous research into logical subsections.

Other secondary sources include literature reviews within previous dissertations and encyclopaedias. Previous dissertations are done at a range of levels from undergraduate degree up to PhD and DSc. PhD theses will have undergone a thorough examination process and if any changes were necessary in the write up as a condition of a PhD award being made, then these changes will have been implemented. Some universities have similar regulations for Master's level dissertations that will eventually be placed in the library. However, undergraduate dissertations and many Master's dissertations will not have undergone such a peer review process before being placed in the library. Therefore, students need to be aware that these literature reviews may not be examples of best practice. Students can pass their research project even if the literature review section receives less than 40 per cent of the marks allocated for that chapter. Students are encouraged to treat such dissertations as secondary sources that identify other material that they should find and read. Literature reviews in previous dissertations may also give students ideas of how to structure their own literature review sections, but the final decision should be based on a more logical consideration of the material reviewed and the different broad areas that have been covered. There are general encyclopaedias and specialised encyclopaedias as described by Thomas *et al.* (2005: 32). While there is no specialised encyclopaedia in performance analysis of sport, there may be some useful definitions of terms that can be obtained from the sports science encyclopaedias identified by Thomas *et al.* (2005).

The ultimate aim of consulting secondary sources is to identify primary sources of academic literature that the student can review. Primary sources are papers describing studies where data have been analysed and findings presented. There are other means of identifying primary sources including internet searches, entering key words into electronic library catalogue searches, lists of abstracts or even just titles of papers presented at research conferences. Internet searches can identify primary sources published in reputable scientific research journals. However, internet searches can also lead to sites containing writings about a topic that have not been peer reviewed and may have been written by authors without any professional training or qualification in the area. These internet sites may be about coaching, sport and sports performance but may have been established and maintained by people purely as a hobby or personal interest. Students should not use material from such sites as primary sources of background material. Other sites may be media sites containing useful quotes that the student may wish to use in initially setting the scene for their own research. The student should be aware that such material is often written in journalistic language that may not be scientific and often sensationalises issues. There are official internet sites of scientific and professional organisations in sport and sports science. Material on these is useful, but the objectivity of such information should always be questioned, particularly when it comes from commercial organisations that are marketing their products and services.

The contents of research conferences can be identified from conference programmes specified on conference websites or in journals that occasionally devote an issue (or a supplement of an issue) to the abstracts from recent research conferences. For example, volume 25 number 3 of the *Journal of Sports Sciences* (February 2007) contains the abstracts of the annual BASES (British Association of Sport and Exercise Sciences) conference that was held in Wolverhampton in September 2006. Performance analysis abstracts can be found within the biomechanics and interdisciplinary sections of such an issue, with other abstracts related to performance analysis being included in the talent development, physiology and psychology sections. *Research Quarterly for Exercise and Sport* produces a supplement to its first issue each year that contains the abstracts of the AAHPERD (American Alliance for Health, Physical Education, Recreation and Dance) annual convention and exposition. Volume 80 number 1 of this journal, published in March 2009, contained the abstracts of the AAHPERD national convention and exposition held in Tampa, Florida in 2008.

Primary sources are identified by secondary sources, internet searches and other means and can be reviewed by the students. These primary sources are original research contributions that describe particular investigations that have been done where conclusions are drawn based on analysis of data. These primary sources refer to previous research when establishing a rationale for the study described. Sometimes, when describing the general background to a study, a primary source may even refer to a secondary source that the student had not previously been aware of. The process of reading secondary and primary sources and identifying further sources of literature continues until what Gratton and Jones (2004: 59) describe as 'saturation', where new sources being read are not identifying any useful sources that have not already been identified during the literature search. Students will notice a diminishing return of new sources being identified, the further they progress with their literature search.

Sources of performance analysis research

Performance analysis of sport has its own specialist journal and there are four key textbooks in the area. The performance analysis community have met every two years or so at their series of World Congresses where research is presented and discussed. The proceedings of these World Congresses have been written up as books, allowing performance analysis students to rapidly survey a great volume of previous research. There are other journals, books and conferences that include performance analysis contributions and the main sources of performance analysis material is listed as follows:

- Performance analysis journal: *International Journal of Performance Analysis of Sport.*
- Other journals that publish performance analysis research: *Journal of Science and Medicine in Sport, International Journal of Sports Science and Coaching, International Journal of Coaching Science, Research Quarterly for Exercise and Sport, Journal of Sports Science and Medicine, International Journal of Sports Medicine and Physical Fitness, Strength and Conditioning Journal, International Journal of Computer Science in Sport, Open Sports Science Journal, European Journal of*

Sports Science, Journal of Human Movement Studies, Journal of Sports Sciences and *Ergonomics*.

- Proceedings of the World Congresses of Performance Analysis of Sport, which are published as books.
- Notational analysis and performance analysis textbooks (Carling *et al.*, 2005, Hughes and Franks, 1997, 2004a, 2008).
- Match analysis and biomechanics sections of Proceedings of World Congresses of Science and Racket Sports and Science and Football, which are published as books.

Indirectly relevant material

Performance analysis of sport is an emerging discipline within sports science, which itself is a relatively new science. Therefore, there will be occasions where students are investigating important and interesting research questions that have not been addressed before in any sport. Students addressing original problems will have difficulty finding directly relevant academic literature to use within the literature review and discussion chapters of their dissertations. In such situations, it is recommended that students draw on indirectly related academic literature. An example of this from the author's own experience was when a colleague and the author came up with the idea to investigate the influence of score line on work-rate in FA Premier League soccer (O'Donoghue and Tenga, 2001). There was an abundance of academic literature on the physical demands of intermittent high intensity activity in field games in general as well as on the demands of soccer match play in particular. There was also a great deal of previous research into time-motion analysis methods as an indirect observational way of estimating the demands of field games. However, the researchers were unable to find any literature on the effect of score line on work-rate in any sport. Indeed, the authors were not able to find any previous research on the effect of score line on any aspect of performance in any sport.

The authors considered what score line at a given point in a match was; they a considered score line to be a reflection of the performance up to that point in the match. This allowed the authors to draw upon literature relating to performance accomplishment. Bandura (1977) stated that performance accomplishments were the most powerful source of efficacy expectations. A goal scored by a team is a performance accomplishment within a match that may, therefore, increase efficacy expectation for the remainder of the match. The researchers also considered possible mechanisms by which work-rate could be affected by score line. If a player attributed the cause of the current game state (especially if the player's team was losing) to factors outside the player's own control, the player's motivation may reduce. This speculation led the researchers to review sports psychology literature on causal attribution and motivation (McAuley, 1992, Weiner, 1985). The initial study done by O'Donoghue and Tenga (2001) found an effect of score line on the percentage of time spent performing high intensity activity, but left the mechanisms involved as a 'black box' that future research could investigate.

This approach used by O'Donoghue and Tenga (2001) is recommended to students who have difficulty finding previously published research that is directly relevant to their own research question. Students should consider the concepts involved in their research question and what general concepts within other disciplines of sports science might be related to them. Students may then identify literature from coaching science sources, sports psychology, biomechanics and exercise physiology that can be discussed in providing a rationale for the research question.

Reading the literature

As has already been mentioned, this stage overlaps with other activities within the literature review and the student may read and record papers while there are still other articles to be identified during the literature search. When reviewing previous literature, the student should be as critical as possible, identifying strengths and weaknesses of previous research. This involves the following steps.

1. Read the title and the abstract to gain an overall synopsis of the purpose of the paper, what was done, what was found and what was concluded.
2. If the paper is of interest to the student's research, then the reference details of the paper should be recorded so the student is able to include the paper in their own reference list. The student may not be able to make a decision about the relevance and usefulness of the paper to their own study after reading the abstract, so this step could occur later.
3. The need for the study as expressed in the introduction or background section should be considered.
4. The student should examine the stated hypotheses that outline the specific research question being answered one way or the other by the paper. Often papers do not formally state hypotheses but instead set out the aims and purpose of the paper. No matter which style is used, the student should initially try to determine the research question being asked and the key variables involved.
5. The student should look at what the study found by examining the results section. The results may agree with or contradict those of other studies that the student has read. The results may be consistent with some theories but not others. These observations should be noted by the student so they are not only considering the paper in isolation, but the review of literature begins to identify patterns and debates within the research area.

6. The student should then consider the methods used within the research described in the paper that led to the results obtained. This order of tasks may seem strange to some, but based on the author's experience in different situations where evidence is considered, very often the findings are understood before the strengths and weaknesses of the methods used are delved into to examine the quality of the evidence.

   The student should first consider the matches, performances and participants included. How many matches were analysed? What level were the matches played at? Were complete or partial matches used? How many matches were used for each team involved? What steps were taken by the researchers to ensure that the matches were representative of the teams or players? Can the performances of the particular teams and players be generalised? Are the matches recent performances or are the data dated to such an extent that rule changes or developments in preparation may have rendered the data irrelevant to the current game?

   Secondly, the student should consider the variables used and the methods used to collect the data. What do the variables actually mean? What was the measurement procedure? Is the measurement procedure, whether observational or otherwise, clear? Is the study replicable? Are there any inherent limitations of the variables as defined by the researchers? Are the variables valid and important aspects of the area of the game being studied? Are the variables measured reliably? Is the level of reliability sufficient in relation to the analytical goals of the study? Have variables been adjusted to address different lengths of performances in sports like tennis? Are the independent variables objective? For example, variables such as positional role may be problematic in some sports where players engage in multiple roles throughout the match.

   Thirdly, the student should consider the data analysis techniques used. Are the statistical procedures the most appropriate given the design of the study? Where independent samples are assumed, are the samples truly independent? Where one variable is used as an independent variable and another variable is used as a dependent variable in a statistical test, is the dependent variable influenced by the independent variable?

7. Once the methods and results have been critically reviewed by the student, the discussion section of the paper should be considered. Have the results been explained in terms of relevant theory and related evidence from previous research? Are there theoretical concepts that could have been discussed in relation to the results that were not included? Is there contradictory evidence from previous research that the paper has failed to acknowledge? Is there supporting evidence from previous research that the paper has failed to recognise? Have any methodological problems affecting the findings and subsequent discussion been acknowledged?

8. The conclusions should then be examined to make sure that these are supported by the findings of the study. The recommendations made should be evidenced by the study. For example, a time-motion study can recommend that those developing conditioning programmes for the given sport are aware of the demands of the sport as expressed in the paper, but it cannot recommend a training programme based on the findings as the effectiveness of such a programme has not been tested by the research.

This process of critically reviewing a paper will be done more comprehensively for some papers than others. The amount of detail used to cover a paper depends on the relevance of the paper to the student's own research question.

Writing the literature review

The final chapter of this book describes the process of writing up research, including writing up the literature review of a dissertation; this information is also briefly summarised here. The student will have a notebook and possibly a word-processed file full of notes made on previous research. This is unstructured in its current form and so the student must decide on a structure of headings and subheadings to use within the literature review. This skeletal outline should be finalised and then the student's notes should be read, highlighting which aspects of previous research to include in which section of the literature review. Before starting to write any of the sections, the student should consider their own research problem and how previous research relates to it. The student is seeking to make an argument justifying the need for their own study based on previous research, therefore the student should aim to critically review previous literature within the structure of this argument rather than reviewing each research paper in turn. When writing a particular section of the literature review, the researcher should extract material from their notebook and mark it as having been used in the literature review. The note form information should be transformed into more scientific writing in contributing to elements of the argument that leads up to a gap in our knowledge of the research topic being identified.

SUMMARY

This chapter has described how students should select a research question and review literature related to that research problem. The criteria for selecting a research problem include the interests, strengths and weaknesses of the student, career aspirations of the student, the importance and relevance of the research problem and the support that can be provided by staff within the student's university department. Some of these criteria may be in conflict and the student often has a difficult decision to make between alternative projects they are considering. Feasibility of undertaking research in the selected area is very important and needs to be considered honestly and fully by the student as early as possible. Once an important, interesting and feasible research problem is selected, the student should review the available literature in that research area. The process of using secondary sources to identify primary sources is not always possible in a young discipline like performance analysis of sport; indeed it is often difficult to identify primary sources of literature related to the research problem. However, the student should endeavour to identify and review theory that is indirectly related to their problem area if necessary.

CHAPTER 4

SPECIFYING THE RESEARCH QUESTION

INTRODUCTION

Once a research topic is selected, it often has to be presented in the form of a research proposal. The research proposal can be the introduction chapter in a thesis or a proposal for future research that has been planned. In quantitative performance analysis research this requires formulating hypotheses in terms of defined variables. The presentation of the research problem also includes the scope of the study and any limitations that need to be acknowledged. The different forms of specification of the research problem serve different purposes. In a dissertation that has been completed, the specification is a statement of what the research problem was. Similarly, an original research paper published in an academic journal will specify the aim and purpose of the study being described. There are other occasions where the research proposal is produced and formally presented before any data collection and analysis takes place. Research proposals can be done at a range of levels from undergraduate student course works to experienced research professionals bidding for research funding. In many universities, the assessment of research methods can include a research proposal for the level 3 dissertation. This research proposal can be done in level 2 allowing the student to identify a research project, apply for ethical approval and allow supervision arrangements to be made.

Students should make every effort to produce a research proposal for a project that is important and interesting to them. If they do not decide what they will study as a research project by the final year of their programme, they may not end up with the most suitable supervisor. The student should also ensure that the research problem is specified as precisely as possible so that the relevant ethics committee can clear the project to commence data collection. If the project specification is incomplete or vague, the ethics committee may not be able to grant ethical approval as the committee will not have been provided with sufficient detail to allow it to make a decision. Clearance to commence data collection could be delayed until half-way through the first semester of the student's final year, a situation that the student should seek to avoid. Research proposals made by academic or commercial researchers when applying for research funding must provide details relating to the purpose of the study, methods, plans, staff costs, equipment costs and consumable costs. The decision of a funding body whether or not to award a research grant depends on cost as well as potential benefits and so the researchers must provide the necessary detail to allow this decision to be made.

The documentation produced with the specification of the research problem usually includes background information and a rationale for the study. This background information and the rationale would have been developed as described in Chapter 3. Therefore, this chapter is primarily concerned with the process of specifying the research question to be answered by the student's research project. As shown in Figure 2.4, this is concerned with 'what' will be investigated and is linked to the format of 'what' will eventually be found by the research.

No matter what form the problem specification takes, a research area will have been selected through various means and motivations including a review of relevant literature. The research area and the rationale for the student's own investigation are transformed into a more precise specification of the research question using the steps shown in Figure 4.1. There are four steps and the remainder of this chapter describes these steps. The example of gender and surface effect on elite tennis strategy (O'Donoghue and Ingram, 2001) is used as an example of specifying a research question.

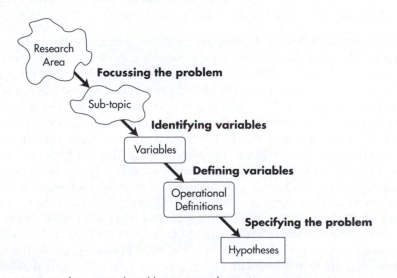

Figure 4.1 Moving from research problem to research question

FOCUSING THE RESEARCH QUESTION

Figure 4.1 shows that the process of specifying the research question gradually transforms the student's chosen research area into a focused and precise research question. The shapes in Figure 4.1 represent material being developed during the process and the arrows represent the four stages that transform the material during this process. The research area is chosen based on the criteria described in Chapter 3 as well as a review of literature in the research area. This is the starting point of the process of specifying a research question.

The first thing that the student must do is restrict the scope of the study so it is manageable. The term 'delimitations' is sometimes used instead of 'scope' within the introduction sections of research projects. Delimitations are restrictions on the study that are set by the researcher. Limitations of the study, on the other hand, are factors that affect the study that are outside the control of the researcher. Limitations will be discussed in the section entitled 'Completing the problem specification', below. An example of restricting the scope of a study is the decisions made when focusing a tennis strategy study (O'Donoghue and Ingram, 2001):

> The study would be restricted to Grand Slam singles tennis from 1997 to 1999; the thesis was submitted in May 1999 and so data were used from the French Open in 1997 to the Australian Open in 1999, allowing the most recent data possible to be gathered, analysed and discussed.
>
> The study was restricted to elite performances at the given tournaments, using the definition of Hughes and Clarke (1995) that a player is considered to be elite on the given surface if they have beaten two fellow professional players at the given tournament. Therefore, only matches from the third round to the final were to be used.

Restricting the scope of the investigation to a particular level of play or type of player is one of the ways in which a problem can be focused (Gratton and Jones, 2004: 40–42); the authors also described temporal, geographical, context and methodological ways of restricting the scope of a research project. Temporal considerations are where the scope of a research project is restricted to a particular era, season or even part of a season. The changing nature of tennis, the physical characteristics of successful players, changes in the way players prepare for competition and rule changes are all factors that necessitate up-to-date research. Therefore, the ball changes that occurred in professional tennis in 2002, and the Australian Open starting to use the same surface as the US Open from 2009, have rendered O'Donoghue and Ingram's (2001) study out of date. Most investigations use up-to-date data and restrict the scope of the study to the current era as a result of this. However, some historical research projects might focus on previous eras as a historical research application of performance analysis.

Setting a geographical scope of a project is useful where little scientific knowledge is available for the given sport in a particular region. This is not to be confused with the use of a cluster sample that is chosen to be representative of some wider geographical area. The region could be one of interest to the student, the local area of the university the student is at or the student's home region. When the research restricts the scope of a project to a particular geographical region, the results are not intended to be representative of the sport outside that geographical region.

The theoretical context of a study may be delimited to a particular existing theory that is being tested. For example, O'Donoghue (2009b) studied service strategy of male tennis players at the French Open, specifically to test aspects of his Interacting Performances Theory (O'Donoghue, 2009a). The study may also be restricted from the outset to using methods that are different to methods used in previous research investigating the problem. These methods could be complementary to previously used methods or the project could be making a methodological advance on previous research. For example, Di Salvo *et al.* (2009) undertook a time-motion analysis investigation of English FA Premier League soccer match play using the ProZone3® system that facilitated the investigation of positional and team quality effects using a very large number of players using highly accurate data (n = 563).

STRUCTURING THE PROBLEM

The research problem is structured by identifying the concepts and variables of interest as well as the hypothetical associations between them. This is a process of abstraction where the complexity of the area is first reduced to a set of concepts, then refined into a set of variables that represent those concepts in a concise form. There is an infinite volume of information that can be derived from match videos to represent the performance, but abstraction allows us to gradually reduce this complex picture of performance to a structured 'finger-print' of variables that characterises the performance. Abstraction is typical in the reductionist approach used in quantitative research and often undeniably important information is excluded from the research. This can be viewed as a weakness of the normative paradigm, but it is also an advantage to be able to reduce a complex problem to a simpler structure of concepts that can be investigated empirically.

Consider the example of tennis strategy at Grand Slam tournaments; the author (O'Donoghue and Ingram, 2001) considered three concepts that formed an initial structure for the research problem: gender, court surface and strategy. The links and associations between these were initially developed using informal processes of intuition, experience, authoritative sources and reasoning.

Gender was initially included due to the purpose of the study and the potential practical application of the eventual results. If differences exist in the nature of tennis match play between men's and women's singles, this knowledge could be used by aspiring tennis players and their coaches. Even if similar styles of play exist between men's and women's singles matches at Grand Slam tournaments, this information would still be important.

Court surface was another broad concept within the scope of the research problem. From a young age, it had always fascinated the author that some players performed relatively better on some surfaces than on others. Commentators often discussed court surface properties and the types of play that were encouraged and discouraged on different court surfaces. Some experienced tennis coaches, players, commentators and spectators might consider it obvious that there are longer rallies played more from the baseline on clay surfaces than on grass surfaces. However, in making decisions about how much practice should be devoted to baseline play and net play, the author believed it was important not only to determine if this was the case, but if it was, exactly how much longer are rallies on clay surfaces than on grass and exactly how many more rallies are played from the baseline on clay than on grass.

The strategy adopted by a player is a plan of action that is considered before the match begins and possibly before the tournament begins. Tactics, on the other hand, are moment-to-moment decisions made by players during the match based on the situation they are in and alternative choices of action open to them (Fuller and Alderson, 1990). Strategy and tactics are not directly observable concepts but are, none the less, important concepts of tennis that could be influenced by gender and surface.

Determining the potential links between the concepts of interest can be done through processes of inductive and/or deductive reasoning. Deductive reasoning is where general theory is formed into a hypothesis that can be tested through a systematic observation of the reality. In performance analysis of sport there are few theories, but common beliefs and speculation about general patterns in sports performance can serve as theories to be tested. In the example of tennis, there are commonly held beliefs about the influence of gender and court surface on strategy that existed before the author undertook his research project. Male athletes are considered to be stronger and more powerful than their female counterparts. This generalisation can be the starting point in a deductive reasoning process leading to a research hypothesis relating to net play in tennis, which is considered to be an aspect of strategy. The text in italics illustrates these reasoning processes, with numbers in parentheses used to identify propositions under consideration:

Male and female tennis players have different anthropometric and fitness characteristics (1)

and

Player body type and fitness may influence the power with which ground strokes can be played (2)

therefore

Gender may influence the power with which ground strokes can be played (3)

The next thing that might be reasoned about is the power with which ground strokes are played and what advantages might be gained by a player with powerful ground strokes and what disadvantages might be experienced by the opponent:

Players with more powerful ground strokes place greater pressure on opponent's shot play than players with less powerful ground strokes (4)

and

When a player is under pressure when playing a shot, the player is less likely to successfully play a passing shot or lob when an opponent is at the net (5)

therefore

Players with more powerful ground strokes are less likely to be passed or lobbed by opponents if they are at the net (6)

A deduction can then be made that a player with powerful ground strokes will use situational probability to make a tactical decision to approach the net or stay at the baseline:

Players with more powerful ground strokes are less likely to be passed or lobbed by opponents if they are at the net (6)

and

A player will be encouraged to approach the net if the opponent's chances of playing a successful passing shot or lob are low (7)

therefore

Players with more powerful ground strokes are more likely to approach the net (8)

This general effect of powerful ground strokes on the chance of a player going to the net can be combined with the generalisation that men play more powerful ground strokes than women to make a deduction about a gender effect on how often players approach the net:

> *Gender may influence the power with which ground strokes can be played (3)*

> *and*

> *Players with more powerful ground strokes are more likely to approach the net (8)*

> *therefore*

> *Gender may influence how often players approach the net (9)*

Inductive reasoning, on the other hand, is where specific observations and experience from the real world lead to the forming of hypotheses through logical reasoning. Previous research is used differently in inductive and deductive reasoning. In deductive reasoning, the theory supported by previous research is used to move towards testable hypotheses. In inductive reasoning, the results or specific evidence provided by previous research is used move in the other direction towards a general testable hypothesis. Observations from particular matches played in Grand Slam tournament finals can be used in a process of induction to provide case study evidence of a gender effect on net strategy:

> *There were 55/162 points (34.0%) of the 2008 US Open men's singles final that were net points which was a greater percentage than the 45/168 points (26.8%) of the 2008 US Open women's singles final that were net points (11) Source: US Open (2008)*

> *and*

> *There were 86/347 points (33.3%) of the 2009 Australian Open men's singles final that were net points which was a greater percentage than the 8/86 points (9.3%) of the 2009 Australian Open women's singles final that were net points (10) Source: Australian Open (2009)*

> *and*

> *There were 48/144 points (33.3%) of the 2009 French Open men's singles final that were net points which was a greater percentage than the 27/142 points (19.0%) of the 2009 French Open women's singles final were net points (13) Source: French Open (2009)*

and

There were 106/413 points (25.7%) of the 2009 Wimbledon men's singles final that were net points which was a greater percentage than the 33/157 points (21.0%) of the 2009 Wimbledon women's singles final that were net points (12) Source: Wimbledon (2009)

therefore

At all four Grand Slam tournaments, there were a greater percentage of net points played in the men's singles final than in the women's singles final (14)

An informal survey of colleagues with knowledge of elite tennis or a focus group can identify beliefs about differences between the men's and women's game based on watching tennis matches at the appropriate level. A critical difference between inductive and deductive reasoning is that inductive reasoning directly or indirectly uses specific real world evidence rather than speculative hunches:

A majority of the most well known net players of the 1980s were men (15)

and

A majority of the most well known net players of the 1990s were men (16)

and

A majority of the most well known net players between 2000 and 2009 were men (17)

therefore

A majority of the most well known net players are men (18)

Useful aspects of the problem that do not involve gender can also be elicited from focus group discussions and combined with empirical research evidence. For example, baseline rallies may be longer than net points and authoritative sources may be able to explain why, at any point in a rally, approaching the net will probably result in a rally ending quicker than staying at the baseline:

A rally usually ends one way or another soon after a player approaches the net (19)

therefore

> *Rallies where one or both players go to the net are typically shorter than baseline rallies (20)*

There may be previous research providing evidence that rallies are longer in the women's game than the men's game. This evidence can be combined with differences between baseline rallies and net points that have been determined through the process of induction. This leads to a testable hypothesis that gender has an influence on the proportion of net points played in tennis matches:

> *Rallies where one or both players go to the net are typically shorter than baseline rallies (20)*
>
> *and*
>
> *Rallies are longer in women's singles than in men's singles (21)*
>
> *therefore*
>
> *There may be a greater proportion of net points played in men's singles than in women's singles (9)*

The hypothesis may be created through more than one pathway of inductive reasoning:

> *At all four Grand Slam tournaments, there were a greater percentage of net points played in the men's singles final than in the women's singles final (14)*
>
> *and*
>
> *A majority of the most well known net players are men (18)*
>
> *therefore*
>
> *There may be a greater proportion of net points played in men's singles than in women's singles (9)*

Figure 4.2 summarises the difference between inductive and deductive reasoning. These reasoning processes are often used subconsciously through rapid thought processes without the researcher recording the various steps involved in forming the hypotheses. Indeed, the rationale for the research question to be investigated may be presented though an argument that justifies the reasoning behind the selection of concepts involved.

A similar reasoning process can be used to determine hypotheses between court surface and strategy. These concepts can then be used to form a structure for the research question. Figure 4.3 shows that strategy is hypothesised to be influenced by gender and court surface, therefore any variables that represent

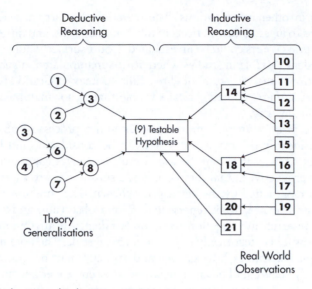

Figure 4.2 Deductive and inductive reasoning

the concept of strategy will be dependent variables. In quantitative research, there are two other types of variable that need to be understood in an example like this: independent variables and categorical variables. In laboratory and field experiments, the effect of some independent variable on some dependent variable is tested. The independent variable is one that is manipulated by the researcher while the dependent variable is one that is tested or observed to measure any experimental effects. The independent variable could be a group membership variable with some participants placed in the experimental treatment group and others placed in a control group. In a cross-over design, all participants would be exposed to experimental and control conditions, with half of the participants being exposed to experimental conditions first and control conditions second, with the other half of the subjects being exposed to these conditions in the reverse order. A correlation study could be undertaken where dosages of the independent variable are manipulated by the researcher to measure the association between the independent variable and the dependent variable. Therefore, the dependent variable can also be thought of as a response variable or a yield variable whose value is hypothesised to be influenced by the independent variable.

There are some variables that are hypothesised to influence some dependent variable that are not manipulated by the researcher, for example gender. These variables are referred to as 'categorical variables' and in our tennis example, gender would be a categorical variable whose effect on the dependent variable(s) relating to strategy is being tested. Court surface is a variable that could potentially be an independent variable that is manipulated by the researchers in some types of study, but a categorical variable that is not manipulated by the researchers in other investigations. The material of court surfaces could be manipulated to investigate the effects of different materials, and different length of grass on grass courts could be investigated. Court surface would certainly be an independent variable in studies where it was manipulated in such ways. In an observational analysis study of elite tennis strategy at Grand Slam tournaments, court surface is a categorical variable that is not manipulated by the researcher in any way.

The concept of strategy is more complex and a process of abstraction is required to identify the key aspects that can be considered, and less critical attributes of strategy that may be too complex to investigate empirically or may be less productive in that they do not add substantially to aspects of strategy already represented by the key aspects chosen. The chosen aspects of the concept of strategy need to be represented in some observable and testable way to allow a research investigation to be undertaken. An attribute of strategy could be measured categorically or on a numerical scale. The potential scale of measurement determines how gender and strategy may be linked within a research study. This could be an association between gender and strategy (two categorical variables) or a difference in the measure of strategy between men's and women's singles tennis (if strategy is represented by some numerical scale variable). The process of abstraction that was used to refine the research problem into the concepts of interest can be used to further refine these concepts into specific variables to be used in the research. The concept of gender is already in the form of a categorical variable and a decision can be made to categorise court surface as well.

The process of identifying particular variables to represent the concept of strategy is not so straightforward. Questions need to be asked about what strategy is. Is strategy a multifaceted concept? If so, what are the different components of strategy? It may also be necessary to consider the limitations of potential methods to be used, as some important aspects of strategy may not be able to be recorded within the observation time available. Strategy in tennis can be made up of the following components:

- service strategy;
- player positioning;

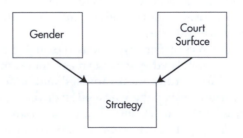

Figure 4.3 Problem structure

- shot placement; and
- net strategy.

Each of these can be further decomposed, through the process of abstraction. For example, service strategy can be based on serve type (slice, flat or kick), pace of serve and placement of serve. Timing factors, such as the duration of a rally, can be also be used as indications of strategy. The percentage of points where an ace is served is a variable that is an abstract representation of serving ability that abstracts away from details of when aces were served, precise service technique, ball placement aspects and receiver positioning. The author (O'Donoghue and Ingram, 2001) used five timing factors, two outcome indicators and nine other variables as a fingerprint representing the concept of strategy:

- mean rally duration;
- mean time between serves played within points where more than one serve was required;
- mean time between points within games;
- mean time between games when the players had to change ends of the court;
- mean time between games when the players did not have to change ends of the court;
- the percentage of points where the first serve was played in;
- the percentage of points won given that the first serve was played in;
- the percentage of points won when a second service was required;
- the percentage of points where an ace was served;
- the percentage of points where a double fault was played;
- the percentage of points where a serve winner was played;
- the percentage of points where a return winner was played;
- the percentage of points where the server approached the net first;
- the percentage of points where the receiver approached the net first;
- the percentage of points that were baseline rallies;
- the percentage of net points won by the player who went to the net first.

DEFINING VARIABLES

There are different types of variables that are used within structured research problems and these can be nominal variables, ordinal variables or variables measured on numerical scales. Nominal variables such as positional role of a player may seem simple enough variables until we start to ask what factors are used in classifying a player as being in one positional role as opposed to another. In netball, players wear bibs that identify their positional role, but in other sports such as soccer different teams use different formations and systems of play, making positional role a complex factor. The concept of 'total football' where players perform multiple roles can make the concept of positional role rather fluid. Positional role can also vary from situation to situation within the same match. The author once analysed the on-field activity of a soccer player who seemed to perform as a left centre back, a left wing back and a left of centre holding midfielder at different points of the same game. On another occasion, the author watched a central defender play as a centre forward when his team required an equalising goal in the last five minutes of a match. Is the switch of position from centre back to centre forward part of the role of the centre back?

In order to be objective, it is necessary to produce an operational definition of positional role so as the classification of player position is independent of the researcher's personal opinion of the player's positional role. Objectivity has other disadvantages and the definition of an elite soccer player used by O'Donoghue (1998) is a good example of this. The definition was that an elite player was an English FA Premier League player with at least one full international cap for his country. The problem with this definition is that some players who have not made any appearances for highly ranked national teams may be better soccer players than some players who have been capped many times by national squads that are not ranked as highly.

Despite the widely held view that notational analysis is a quantitative observational analysis method, a great deal of notational analysis activity involves subjective evaluation of player behaviour. For example, the classification of locomotive movement during time-motion analysis involves observer perception of movement and largely subjective judgements when classifying movement. The point at which a player makes the transition from jogging to running and from running to sprinting will vary from observer to observer. The frequencies of movement instances are simply numerical counts of subjective classifications.

There are, however, occasions where it is possible to define variables in a precise and objective manner. Point type in tennis is used as an example of producing operational definitions. It is often useful to specify some basic terms first. For example, if we start by trying to define a double fault, we might define it as follows:

> *A double fault is served when both the first and second services fail to land in the correct service box.*

The problem with this definition is that it is incomplete and a smart tennis lawyer could argue that a table tennis-type serve could be counted as being in because we did not specify that the first contact of the ball with the court had to be in the correct court on the other side of the net from the server. Specifying all of the elements of a double fault including foot faulting, excluding lets, the ball having to pass over the net to be a good serve and regulations regarding the service action and multiple ball contact of the racket can become very complicated when specifying a double fault due to the fact that there are at least two serves involved. So initially, it is a good idea to define a 'good serve' and a 'service fault'. These terms can then be used within the definition of an ace and a double fault.

The definition of a net point presents similar problems. How far does the player have to travel towards the net before the point is considered to be a net point? O'Donoghue and Ingram (2001) counted a player as having approached the net if the player had crossed the service line (back of the service boxes) and the player or the opponent still had to play one or more shots in the point. So if the player hit a winner from behind the service line and ran towards the net after playing this shot, it was not a net point. If the opponent reaches this shot but plays an error (another term we need to define), but before the player crosses the service line, it is not a net point. However, if the opponent reached the shot after the player has crossed the service line, whether successfully returning the ball or not, the point counts as a net point. If the player played any shots after crossing the service line, then the point is a net point, irrespective of the outcome of these shots.

There are occasions where a single variable may not be enough to characterise the concept of interest. If we have operationally defined the duration of a rally, the mean rally duration for the match can be computed but this will not tell the reader if all of the rallies in matches are consistently around the mean duration or if there is a wide spread of rally durations within matches. The standard deviation of the mean rally duration will simply represent the spread of mean rally durations across matches rather than within matches. A match may have a mean rally duration of 4s with some rallies being longer than 10s. It is not possible that there will be any rallies of negative durations and so rally duration within matches has a skewed distribution of values. Therefore, there is a case for using the median rally duration of a match as a measure of average rally duration. However, the median will not provide an indication of the spread of rally durations within matches. O'Donoghue and Liddle (1998) used a series of variables to represent the percentage of points where the rallies were in different 2s duration bands (less than 2s, 2s – under 4s, 4s – under 6s, 6s – under 8s, 8s – under 10s and so on). This not only allowed the modal rally duration to be determined but also allowed the distribution of rally durations within matches to be studied.

FORMING HYPOTHESES

To use hypotheses or not?

The use of operational definitions and formally expressed hypotheses can add considerably to the word count of the introduction of a dissertation. When one considers the number of words in a mobile phone contract, it becomes apparent that a large number of words are required to specify any-thing completely and unambiguously. Therefore, some research projects will use a less precise set of aims and a statement of the purpose within the introduction. Another reason for not expressing formal hypotheses is that the student may be producing an initial proposal for a project and part of the research project is to identify performance indicators and develop a system to allow these to be determined. When such a student eventually completes their dissertation, the operational definitions may appear in the methods section rather than in the introduction. Other students may specify some fundamental terms in the introduction and define other terms within the methods chapter.

There is a more fundamental reason why formal hypotheses may not be specified in some projects. There are some research projects that simply describe the performance of a group of interest using a single sample and descriptive statistics. Inferential statistics are not relevant to the purpose of such studies unless values are being compared with some benchmark standard. Therefore, research projects can be done using a single sample that is described with-out any inferential testing and hence without any need to specify formal hypotheses.

The remaining sections on hypotheses are relevant when formal hypotheses are to be used. They describe how to specify hypotheses when different types of variable are being used. A hypothesis is a precisely specified outcome of a study or experiment. In Figure 4.1, the formulation of hypotheses is done once an important research question of interest has been decided, structured and operationalised. The hypotheses specify precisely the research question that the research project seeks to answer. When undertaking a quantitative study, the hypotheses are the possible outcomes of the study, and the research question becomes one of determining which outcome to accept and which to reject. Typically there are two hypotheses: a null hypothesis (H_0) and an alternative hypothesis (H_A), which are mutually exclusive as well as being the only possi-ble outcomes of the study. The alternative hypothesis is sometimes referred to as the 'research hypothesis' because it is the link or association between con-cepts suggested during the structuring of the research problem. The null hypothesis is a potential outcome reflecting no association or link between the concepts of interest. 'Mutually exclusive' means that there is no other outcome of the study that can be interpreted as agreeing with both or neither hypothesis. The hypotheses must be observable or testable so that they can be accepted or refuted based on the evidence of the data collected. Well formed hypotheses are also an excellent starting point when designing the methods to be used.

Hypotheses involving only nominal variables

If we wished to determine if the court surface had an influence on the chances of an upset in Grand Slam singles tennis, we would be dealing with two categorical variables that just happen to be nominal variables. The specification of the research question in this case must not attempt to impose tests of order on either variable. Logically we do have a hypothesised dependent variable – match outcome, which is a dichotomous variable with values being 'a win for the higher ranked player' or 'an upset' with respect to the World rankings used in professional tennis. This variable is hypothesised to be the dependent variable rather than court surface because the court will not turn from cement to grass even if the World number 1 is defeated by the World number 500! The study may be surveying matches played on two or more court surfaces. If more than two court surfaces are included in the study, the researcher may wish to know which pairs of court surfaces are different if there is a surface effect. However, this is best described in the methods and the eventual results. The hypotheses will be whether there is a surface effect on match outcome or not.

In the following examples of hypotheses, it is assumed that the scope of the investigation has been restricted to singles tennis matches played at Grand Slam tennis tournaments. The following are examples of poorly presented hypotheses:

H_0 – *Court surface has a small influence on match outcome.*
H_A – *Court surface has a large influence on match outcome.*

These null and alternative hypotheses do not cover every possible outcome of the study. This is because there may be no influence at all of court surface on match outcome, with exactly the same proportion of upsets being observed on all court surfaces. Furthermore, the distinction between a small and large influence may not have been defined. The following hypotheses are vague if match outcome has not been defined:

H_0 – *Court surface has no influence on match outcome.*
H_A – *Court surface has an influence on match outcome.*

Assuming that an 'upset' has been defined as any tennis match won by the lower ranked player according to the World rankings, then the hypotheses can be expressed satisfactorily as follows:

H_0 – *Court surface has no influence on the proportion of matches that are upsets.*
H_A – *Court surface has an influence on the proportion of matches that are upsets.*

If there were only two court surfaces included in the study (for example clay and grass), the hypotheses might be expressed as follows, which would be problematic. The problem here is that the alternative hypothesis is one-tailed assuming a direction of any difference: that if there is a difference it is that there is a greater proportion of upsets on grass. One tailed assumptions can be used, but only when there is strong theoretical grounds to be assuming the direction of any difference found. The hypotheses below introduce an additional problem in that they do not include the possible outcome that there may be a higher proportion of upsets on clay:

H_0 – *The proportion of matches that are upsets is similar between matches played on grass and clay courts.*
H_A – *A greater proportion of matches are upsets on grass surfaces than on clay surfaces.*

A two-tailed version of the alternative hypothesis is shown below where the outcome is a difference in the proportion of outcomes observed without any assumption about which the surface is associated with more upsets if there is a surface effect:

H_0 – *The proportion of matches that are upsets is similar between matches played on grass and clay courts.*
H_A – *The proportion of matches that are upsets varies between matches played on grass and clay courts.*

At the moment, these hypotheses could be used for a study of just men's singles matches, or a study of just women's singles matches, or a study that combines both. However, with both men's and women's matches involved, it would be a good idea to include gender as another categorical factor that might have an influence on the number of upsets. This could be done by having two sets of hypotheses as follows:

$H_0 1$ – *Court surface has no influence on the proportion of matches that are upsets.*
$H_0 2$ – *The proportion of matches that are upsets is similar between men's and women's singles matches.*
$H_A 1$ – *Court surface has an influence on the proportion of matches that are upsets.*
$H_A 2$ – *The proportion of matches that are upsets differs between men's and women's singles matches.*

This example looks at the influence of gender on match outcome and the influence of court surface on match outcome separately. The question about the effect of court surface involves both men's and women's singles matches pooled together. Similarly, the question about the effect of gender on match outcome involves matches played on different surfaces. The readers of the eventual study might like to know whether the surface effect occurs in men's singles and women's singles when considered separately. This would require the hypotheses to be worded as follows:

H_0 – Court surface has no influence on the proportion of matches that are upsets in men's or women's singles tennis.
H_A – Court surface has an influence on the proportion of matches that are upsets in men's singles tennis and/or women's singles tennis.

Note that the use of the 'and/or' in the alternative hypothesis is necessitated by the need to ensure that the alternative hypothesis is the exact opposite outcome to the null hypothesis. Therefore, if the eventual results show a significant influence of court surface on match outcome in men's singles, women's singles, or both men's and women's singles tennis then the null hypothesis can be rejected.

Hypotheses involving only numerical variables

There are occasions where the research question is about the association or relationship between two or more numerical scale variables. In such cases the hypotheses are not expressed in terms of samples, groups or conditions unless some additional categorical variable is included. Instead, the alternative (research) hypothesis is that there is some relationship between the numerical variables.

Consider the example of serving in Grand Slam singles tennis where a researcher anticipates that if the first serve is played fast and close to the lines of the service box, it will be out more often than a slow first serve aimed well inside the target service box. However, on those occasions where such a fast serve is successfully played into the court, it will be more likely to lead to a winning point for the server than if a slow service aimed well inside the target service box were used. This reasoning could justify the use of a one-tailed hypothesis that assumes that any association between the percentage of first serves that are played in and the percentage of points won by the server must be a negative relationship. A negative relationship between two numerical variables X and Y is one where as X increases, Y decreases. The hypotheses in this tennis serving example might be expressed as follows:

H_0 – There is a positive relationship or no relationship between the percentage of points where the first serve is in and the percentage of points that are won by the server, given that the first serve is played in.
H_A – There is a negative relationship between the percentage of points where the first serve is in and the percentage of points that are won by the server, given that the first serve is played in.

A more cautious researcher might express these hypotheses in a two-tailed manner, not assuming a particular direction of any relationship that might be found between the variables:

H_0 – *There is no relationship between the percentage of points where the first serve is in and the percentage of points that are won by the server, given that the first serve is played in.*

H_A – *There is a relationship between the percentage of points where the first serve is in and the percentage of points that are won by the server, given that the first serve is played in.*

If we wanted to look at this relationship for the two genders separately, the hypotheses could be presented as follows:

H_0 – *There is no relationship between the percentage of points where the first serve is in and the percentage of points that are won by the server, given that the first serve is played in for men's singles matches or women's singles matches.*

H_A – *There is a relationship between the percentage of points where the first serve is in and the percentage of points that are won by the server, given that the first serve is played in for men's singles matches and/or women's singles matches.*

If we wanted to investigate this relationship between the percentage of first serves played in and the percentage of points won, given that the first serve is played in for the two genders and four court surfaces separately (that is eight conditions), then the hypotheses would be expressed as follows:

H_0 – *There is no relationship between the percentage of points where the first serve is in and the percentage of points that are won by the server, given that the first serve is played in for men's singles matches or women's singles matches on any court surface in Grand Slam tennis.*

H_A – *There is a relationship between the percentage of points where the first serve is in and the percentage of points that are won by the server, given that the first serve is played in for men's singles matches and/or women's singles matches on one or more court surfaces used at Grand Slam tournaments.*

It should be noted that if a relationship is found between the percentage of points where the first serve is played in and the percentage of points won by the server, given that the first serve is in, on one or more court surfaces in men's or women's singles tennis, then the null hypothesis can be rejected. The hypotheses for this example still leave some methodological questions. For example, when gathering data, should the service points for the two players in a match be considered together or as separate player serving performances? If considered together, how should the two performances be combined? One player may have played more service points than the other, so deriving the two variables from the total number of first serves in the match, the total number of points and the total number of points emanating from a first serve that are won by the server may weight one player's performance too highly within the data. Therefore, the two variables of interest could be computed separately for the two players within a match and then combined by determining the mean of the value for each variable within the match.

The introduction of categorical factors such as gender and court surface helps to increase the volume of potential results that can be produced from the research project. This is something that should be considered at this stage of the project. The 'V' shaped diagram in Figure 2.1 shows the link between hypotheses and results, and is intended to make the point that the results format should be considered before finalising the specification of the research question. It is very rare that a research project would be designed in a way that only performed one correlation test between two variables, no matter how labour-intensive data collection is. The results in such a rare case would consist of a single scatter graph and a coefficient of correlation. It is more likely that a study would explore associations between multiple variables.

Consider what may happen if the example of tennis serving were extended to include two other numerical scale variables: the percentage of second serves that are in and the percentage of points won, given that the second serve is played in. The following hypotheses are an example of a lazy researcher's specification of hypotheses. Without any theoretical justification, the researcher simply states that they will test to see if there are any correlations between any pair of the four variables:

> H_0 – There is no relationship between any pair of the following varia-
> bles: a) the percentage of points where the first serve is in, b) the per-
> centage of second serve points where the second serve is in, c) the
> percentage of points that are won by the server given that the first serve
> is played in. and d) the percentage of points that are won by the server
> given that the second serve is played in.
>
> H_A – There is a relationship between one or more pairs of the following
> variables: a) the percentage of points where the first serve is in, b) the
> percentage of second serve points where the second serve is in, c) the
> percentage of points that are won by the server given that the first serve
> is played in, and d) the percentage of points that are won by the server
> given that the second serve is played in.

In this example, it is not recommended that the researcher specifies an alternative hypothesis that there is a relationship between one or more pair of variables. There may be other examples where such hypotheses are suitable, but in this example, it shows lack of direction from theory, logic and reasoning. One might reason that the percentage of points won on first (or second) serve could be negatively related to the percentage of first (or second) serves that are played in. Similarly, one might reason that there might be a relationship between the percentage of first serves played in and the percentage of second serves played in due to the serving accuracy developed by the player. The percentage of points won when the first serve is in and when a second serve is in could be related due to player ability when playing ground strokes, volleys, positioning, movement and other aspects of play that follow the service. The potential link between the percentage of first serves that are played in and the percentage of points won, given that a second serve is played in, may be deemed as too indirect to warrant inclusion within the hypotheses stated. This would result in the hypotheses being expressed as follows:

> $H_0 1$ – *There is no relationship between the percentage of first serves that are played in and the percentage of second serves that are played in.*
> $H_0 2$ – *There is no relationship between the percentage of first serves that are played in and the percentage of points won, given that the first serve is played in.*
> $H_0 3$ – *There is no relationship between the percentage of second serves that are played in and the percentage of points won, given that the second serve is played in.*
> $H_0 4$ – *There is no relationship between the percentage of points won, given that the first serve is played in and the percentage of points won, given that a second serve is required and is played in.*
> $H_A 1$ – *There is a relationship between the percentage of first serves that are played in and the percentage of second serves that are played in.*
> $H_A 2$ – *There is a relationship between the percentage of first serves that are played in and the percentage of points won, given that the first serve is played in.*
> $H_A 3$ – *There is a relationship between the percentage of second serves that are played in and the percentage of points won, given that the second serve is played in.*
> $H_A 4$ – *There is a relationship between the percentage of points won, given that the first serve is played in and the percentage of points won, given that a second serve is required and is played in.*

Some relations are linear while others are curvilinear. Some associations may be positive or negative while others might show an optimal sub-range of some X variable that maximises a Y variable. These details of correlation types are best left to the methods section, with the hypotheses simply expressing that a relationship is being tested for.

Hypotheses involving categorical factors and numerical dependent variables

The ease or difficulty of writing hypotheses depends on the number of variables involved. Hypotheses about the influence of some categorical variables (such as gender or court surface) on some numerical scale variables (such as mean rally duration) can become complex if there are multiple categorical factors and/or multiple hypothesised dependent variables.

The following hypotheses are an example of poor practice because the reader will not know what is meant by 'rallies'. Are we talking about shots played? Are we talking about rally duration? Are we talking about strategic or technical aspects of rallies?

H_0 – *There is no difference in rallies between men's and women's singles matches.*
H_A – *There is a difference in rallies between men's and women's singles matches.*

If using rally times as an indication of strategy, it is better to use the term 'rally duration' rather than 'rally length' because rally length could be referring to the number of shots played or the duration (in s) of the rally. Assuming that rally duration is defined as the time between start of a rally and the end of a rally, we would still need to define the points in time where a rally starts and ends. The start of the rally may be defined as when the serving player's serve strikes the ball with the serve being a 'good serve' as opposed to a 'service fault'. The end of the rally could be defined as when the ball hits the net, lands out of court, or bounces twice before being played by one of the players. By this definition, double faults do not count as rallies but aces do. The 'rally' where an ace is served ends when the served ball bounces for a second time without the receiving player reaching it. Despite definitions of rally starts and ends, service faults, good serves, winners and errors, the following hypotheses are still unsatisfactory. The problem with these hypotheses is that the reader will not understand what we mean by 'rally duration' because there are lots of rallies in a match. Are we talking about the longest rally? Are we talking about the mean rally duration? Are we talking about the median rally duration?

H_0 – *There is no difference in rally duration between men's and women's singles matches.*
H_A – *There is a difference in rally duration between men's and women's singles matches.*

'Mean rally duration' is one term that can overcome this problem. However, the one-tailed hypotheses below assume a particular direction of a difference between men's and women's singles tennis:

H_0 – *The mean rally duration in men's singles matches is shorter than in women's singles matches.*
H_A – *The mean rally duration in men's singles matches is not shorter than in women's singles matches.*

Without a sound theoretical basis for assuming that rallies in men's singles might be shorter than in women's singles, the two-tailed hypotheses below are used:

H_0 – *There is no difference in mean rally duration between men's and women's singles matches.*
H_A – *There is a difference in mean rally duration between men's and women's singles matches.*

If another factor such as court surface is introduced, the hypotheses may be expressed as follows:

H_0 – *Neither gender, court surface nor the interaction of gender and court surface have an influence on mean rally duration.*
H_A – *One or more of gender, court surface and the interaction of gender and court surface have an influence on mean rally duration.*

Note that these hypotheses include the interaction of the two factors (gender and court surface), which may not be able to be tested if parametric procedures are ruled out. This is a decision that cannot be made until the data are explored, but it does not prevent us from stating the hypotheses we would wish to test if the data satisfy the assumptions of parametric procedures. Furthermore, post hoc hypotheses are not specified, as differences between pairs of tournaments are particular manifestations of the alternative hypothesis, which can be dealt with during the methods and results chapters.

The study undertaken by the author (O'Donoghue and Ingram, 2001) had several variables that were potential indicators of strategy that the author wished to compare between men's and women's singles and between tennis matches played on different court surfaces; these variables are listed on page 89. This large set of performance indicators could be determined for each match by the software system developed to capture and manage tennis data during observation of match videos. The hypotheses of the investigation could, therefore, be expressed as follows:

H_0 – *Neither gender, court surface nor the interaction of gender and court surface have an influence on any of the following:*
Mean rally duration.
Mean time between serves played within points where more than one serve was required.
Mean time between points within games.
Mean time between games when the players had to change ends of the court.
Mean time between games when the players did not have to change ends of the court.
The percentage of points where the first serve was played in.

The percentage of points won given that the first serve was played in.
The percentage of points won when a second service was required.
The percentage of points where an ace was served.
The percentage of points where a double fault was played.
The percentage of points where a serve winner was played.
The percentage of points where a return winner was played.
The percentage of points where the server approached the net first.
The percentage of points where the receiver approached the net first.
The percentage of baseline rallies.
The percentage of net points won by the player who went to the net first.
H_A – *One or more of gender, court surface and the interaction of gender and court surface have an influence on one or more of the variables listed in the null hypothesis.*

It could be that the inclusion of so many dependent variables drastically increases the chance of the null hypothesis being rejected. This does not invalidate the study, and if the researcher is concerned that the null hypothesis might be rejected by chance due to the number of tests being performed, there are statistical means of making tests stricter or using initial tests that contain the experiment-wide sampling error probability. This will be discussed in Chapter 8.

COMPLETING THE PROBLEM SPECIFICATION

Once the student has identified a research area of interest as described in Chapter 3 and gone through the steps required to specify a research question that have been covered in this chapter, the student can complete the introduction chapter of their dissertation. The contents of this chapter will typically include:

- A background to the study that briefly states the problem and justifies the originality, relevance and need for the study.
- The purpose, aims and associated objectives.
- The scope and limitations of the study.
- Definitions of terms.
- Hypotheses.

The limitations of a study are factors that are outside the control of the researcher that may influence the results obtained. These include injuries, fitness of players, opposition effects, weather conditions, importance of the matches being analysed, tactical decisions made, team selection and player preparation for matches. Fuller details of how to write up the introduction are covered in Chapter 10.

SUMMARY

This chapter has described how a vague research problem is transformed into a precise research question through a four-stage process. First, the student focuses the research problem by defining the scope of the study. The second stage involves identifying the key concepts of the research problem and links between these concepts. This structuring process leads to a research question being identified in terms of these concepts. During the third stage, variables are defined to represent the concepts of interest through a process of abstraction. Finally, formal hypotheses can be specified so the research question becomes one of which hypothesis to reject.

Often in performance analysis, the selection and definition of variables is part of the system development process within the project, meaning that formal hypotheses cannot be produced within a research proposal. There are other research projects whose purpose is to describe the performance of a single sample and, therefore, the research will not involve inferential statistical tests and formal hypotheses may not be appropriate. Formal hypotheses would not be used in qualitative research studies. In any of the situations where formal hypotheses cannot be specified, the student should at least state the aim and purpose of the investigation. The specification of the research is completed by providing a brief background to the research problem, a rationale for the study, and specifying the scope and limitations of the research.

CHAPTER 5

ETHICAL ISSUES IN PERFORMANCE ANALYSIS

INTRODUCTION

'Ethics' is a word that has many different meanings in different contexts. In many universities and other higher education institutions, ethics frameworks outline principles for the conduct of staff and students as well as the values of the institution. Institutions expect students to engage in their programme of study, behave appropriately and within regulations, without causing the institution, its partners or the local community harm. Similarly, institutions expect their academic staff to have respect for colleagues, students and other partners, and to behave professionally and responsibly in all areas of their work for the institution and when representing the institution. Staff and students of the institution should act within the law, behave fairly and respect the dignity and rights of others.

In addition to ethics frameworks, universities and other academic institutions have ethics principles, policies and procedures specifically for research activity. It is research ethics that is covered in this chapter, which is broadly divided into two parts. The first part covers scientific dishonesty, based on the seven sources of scientific dishonesty identified by Shore (1991, described by Thomas *et al.*, 2005: 77–85) and considers them specifically within performance analysis research. The second part of the chapter is concerned with the principles, policies and procedures for research ethics. Research ethics principles apply to the way research is carried out and reported. Universities and other higher education institutions have a responsibility to ensure that the research they undertake is done legally, safely and responsibly.

SCIENTIFIC DISHONESTY IN PERFORMANCE ANALYSIS

Thomas *et al.* (2005: 77–85) described the following seven sources of scientific dishonesty listed by Shore:

1. plagiarism;
2. fabrication;
3. omission;
4. faulty data gathering;
5. poor data storage and retention;
6. misleading authorship;
7. unacceptable publication practices.

Plagiarism

Plagiarism of ideas, writings and diagrams is scientific dishonesty that is heavily punished at all universities and academic institutions (Thomas *et al.*, 2005: 77–8). The penalty for plagiarism in some institutions is the discontinuation of the student's programme of study without any possibility to repeat the coursework. The consequences of plagiarism for the individual may be long lasting with damage to their academic reputation and career prospects. Academic staff are obliged to give accurate references for students. Where a student has committed plagiarism and a reference request has been sent to a member of staff asking if the student is honest and trustworthy, the member of staff should accurately report that the student has been dishonest. This may seem severe, but taking credit for someone else's work is a serious matter and universities and other academic institutions need to be fair to all of their students. Students need to be aware not only of these consequences but also that they are very likely to be caught if they have plagiarised material. The Turnitin system (iParadigms, Oakland, CA, USA) is one of the packages used by academic staff to automatically inspect student work, comparing it with a very wide set of published material. Similarities are identified and reported in a way that provides evidence of the plagiarised sections and the alleged source of the material. This evidence can then be used in unfair practice investigations.

A common act of plagiarism by students is where they copy material from published sources, intending to reword it at a later time. Very often due to time pressures and poor organisation, they forget which sections of a large document they are producing that they copied from elsewhere and even forget that they copied it at all. Even where such students manage to reword the copied sections, the practice is to be discouraged. Students should try to read and understand published research to the extent that they can write about it in their own words. This is beneficial to the student's learning and development. Thomas *et al.* (2005: 78) described inadvertent plagiarism, where a co-author of a paper that includes plagiarised material is not the person responsible for including non-original material. This can still have serious consequences for the co-author concerned and, therefore, Thomas *et al.* (2005: 78) recommended that co-authors of papers should thoroughly read the paper before it is submitted to avoid such problems.

Related to plagiarism is the use of copyrighted material. Blatant scanning of copyrighted material (tables and diagrams) in student work will be penalised even if the source is referenced. Students should adapt or reproduce such diagrams or tables, referring to the source of the material.

Fabrication

Fabrication is where quantitative or qualitative data are made up by researchers without actually gathering the data. Some or all of the data within a study may be fabricated and this type of dishonesty may have consequences far beyond the study itself. Published investigations that have used fabricated data may be used by practitioners when making important decisions about match preparation. In performance analysis, the data within a reliability study or within the main investigation may include fabricated values. Reliability and objectivity are major issues in performance analysis as many systems include human operator components. If a method is found to be unreliable, it can render the entire study invalid. There are honest ways of dealing with limited reliability including altering methods to improve the level of reliability, end-user training (O'Donoghue *et al.*, 1996c) and using less precise measurements that may be more reliable (O'Donoghue, 2007b). However, some researchers may choose to be dishonest and fabricate reliability data. One example is reporting intra-operator agreement studies as inter-operator agreement studies where the researcher has not been well enough organised to train additional operators for the purpose of demonstrating reliability. Another example is making a copy of one observer's data and introducing minor errors into the copy in an attempt to make the data appear believable. Such dishonest practice is often easily identified because the results may be too good to be true!

Once the reliability study of a performance analysis research project has been completed, the student can embark on the main study being undertaken, which may not produce significant findings. Researchers should report honestly the findings of their investigations, especially if these are being published and others may make decisions based upon the results. Some students undertaking research projects are under the impression that a study obtaining significant results will be awarded a higher grade than one that does not. Where the influence of some independent factor on dependent performance indicators is being investigated, it will be unknown prior to the investigation whether the factor has a significant influence or not on the hypothesised dependent variables. Therefore, the results of the study, whether statistically significant or not, will be a contribution to knowledge of sports performance. However, some dishonest students may choose to fabricate some data to ensure a significant result. A researcher may do this as significant results of a preliminary study may lead to a greater chance of publication or being awarded external research funding. Both of these achievements will help the career prospects of the individual. However, such pressures must be resisted as fabrication of data in such situations is fraudulent.

Omission of data

The third source of scientific dishonesty ranges from the omission of individual data items that do not suit a research hypothesis to a researcher intentionally choosing not to publish a completed study where the results do not suit some theory being promoted by the researcher. In performance analysis research, there are typically criteria for the inclusion and exclusion of matches and performances. For example, a data set of Grand Slam tennis matches was analysed by O'Donoghue and Brown (2009) to investigate momentum. Only matches of at least 200 points could be included in the investigation, meaning that only 13 men's singles matches out of 75 were used. This type of exclusion of data is reasonable and the criteria for inclusion are transparent. The exclusion of data becomes dishonest where matches that satisfy the criteria for inclusion in the study are not used in the analysis so that the researcher can achieve a desired outcome of the study. Data values could be omitted from both the reliability study and the main study of the research project. In both cases, it is still dishonest and, just like fabrication, can have implications beyond the study where the results are used by others to inform decision making.

In some disciplines of sport and exercise science, it is common to exclude statistical outliers and extreme values. Outliers and extreme values may be considered as being due to measurement error in some studies. However, in other studies these outliers and extreme values may represent legitimate performances. For example, in the data set of previous World Cup soccer matches used by O'Donoghue (2006c) to produce predictive models of the 2006 FIFA World Cup, there were several high scoring matches that were statistical outliers. O'Donoghue (2006c) chose to include these within his modelling data because they were real data values that occurred in actual sports performance and not values resulting from measurement error.

Non-publication of an entire study may be the decision of the authors(s), reviewers for a journal or conference, or the journal editor or chair of the scientific committee of a conference. In any of these cases, research that was actually done and results that were actually found are being concealed from the public. It is recognised that journals, conferences and other outlets have limited capacities for publication and some set higher standards than others. However, there is always the possibility that a reviewer or editor may recommend the rejection of a paper for less than honest reasons. For example, the conclusions of the paper may contradict research done by the reviewer or invalidate some training procedure that the reviewer intends to exploit over the coming years. The reviewer may be doing similar work that they wish to publish before the paper they are reviewing is published. A further possibility is that national research audits on which research funding decisions are made (such as the Research Assessment Exercise in the UK) lead to dishonest practices by academics who may abuse their position on editorial boards and as journal reviewers. Some will argue that blind reviewing is done, but in reality it is not too difficult to determine the authors of papers based on references to other studies within the authors' programme of research. Delaying or stopping research undertaken at competitor universities from being published may assist the reviewer's university in the national research audit. However, such practices are examples of gross scientific dishonesty.

Researchers may do a series of studies over a period of time in order to produce evidence to support the theory they are developing. It is important that any theory is supported by evidence so that the theory is independent of the author's personal opinion and that any evidence contradicting the theory is also acknowledged. To illustrate this point: the author of this book has been concerned about the importance of satisfying the assumptions of statistical modelling techniques. The author wished to determine how serious violating the assumptions of statistical procedures actually was. To investigate this, models of performance outcomes of soccer and rugby matches were produced. Each study contained models where the data used were transformed to satisfy the assumptions of the modelling techniques and models where the data violated those assumptions. The tournaments used in the investigations were the 2003 Rugby World Cup (O'Donoghue and Williams, 2004), Euro 2004 (O'Donoghue, 2005b) and the 2006 FIFA World Cup (O'Donoghue, 2006c). The outcomes of matches were then predicted by the models and compared with the actual outcomes of those matches. All three of these studies concluded that the models based on data that violated the assumptions of the modelling techniques were more accurate predictors of actual sports performance than the models where the data were transformed to satisfy the assumptions. O'Donoghue (2009c) did a study on Euro 2008 that provided contradicting evidence that satisfying the assumptions of the modelling techniques did lead to more accurate predictions of actual sports performance than when the models used data that violated the assumptions. It was a matter of scientific honesty that this study was presented in order to give a balanced and fair view of the evidence for and against the effectiveness of satisfying the assumptions of the modelling techniques.

Faulty data gathering

Faulty data gathering is where the researcher is aware of problems with the methods or systems used during data gathering that may have affected some or all of the data that were collected. This becomes a matter of scientific dishonesty where the researcher reports results based on the analysis of this data without acknowledging data gathering problems. In performance analysis, there are many methods that have limited reliability. This is not a problem where the researchers test their methods and provide reliability results allowing readers to understand the results in the knowledge of limited reliability. There are, however, many possible situations where data are not of the quality portrayed in the methods sections of dissertations and published research papers.

There may be occasions where students use untrained operators to increase the number of performances that they can include in the study. Another example of faulty data gathering is where a known and statistically significant observer bias is neither treated nor acknowledged when writing up the research. Observer bias is where one observer has a known tendency to record higher or lower values for some performance indicator than other observers. Similarly, there may be known problems in the classification of events in early matches before the operation of the system was properly discussed. These matches should not be included in the data analysed.

Poor data storage and retention

Poor data storage and retention is an area of scientific dishonesty that is more associated with research involving human tissue and other material where samples can be affected by storage conditions. There are possible examples of poor data storage that could lead to dishonest practices where researchers analyse data and publish results without acknowledging known problems with the data. Performance analysis research may involve audio cassettes or video cassettes that may stretch (Leger and Rouillard, 1983) or equipment may have differing recording and playback speeds. Video data can be stored in different CODECs on external hard disks and may be required for future use. The conversion process from one file type to another and the need for compression may reduce the quality of the video footage. Researchers are obliged to acknowledge any such problems with the video sequences they are using in their research. Researchers should facilitate traceability from the results of a study to the raw data gathered by securely keeping video recordings in case there is a need to verify results any time in the future.

Another example of a data storage issue is where manual data forms are stored over several seasons and then a slight change of system might divide one type of behaviour into different subclasses. This will mean that frequencies for these new behaviours are only known for matches collected with the suitably revised forms. For example, a netball tallying system may have been used to record possessions and scores from centre passes and turnovers. The system may be revised to distinguish between turnovers in open play and turnovers involving a restart of play. Any attempt by the student or researcher to best guess how the previously combined turnover category might have been distributed between open play turnovers and turnovers involving a restart is fabrication of data.

Misleading authorship

Misleading authorship is a problem in many areas of science. There are conventions on the order of authors in different disciplines. In performance analysis of sport, the order of authors should be in decreasing order of academic contribution to the paper. For a supervisor's name to appear on a paper, the supervisor's contribution must be more than mere encouragement and management of the research. The supervisor must make some academic contribution to warrant their name being included as an author. Researchers should operate a strict policy of:

- no passengers;
- no hijackers; and
- no excess baggage.

This is in the interests of fairness to people's careers so that when applying for positions or promotions, all applicants can be accurately assessed on the quality and quantity of their achievements and capabilities.

In quantitative research, it is possible to use a data gathering instrument with a human element. This typically involves system operators being trained to use the match analysis system to record data without undertaking any data analysis during the process. Some research in performance analysis can use human observers to collect the data whose contribution would not warrant inclusion as an author of the paper but should be mentioned in the acknowledgements. Of course it is advisable to inform human observers being used that they will not be authors of the paper at the outset of the project. Similarly, technician and administrative support does not usually warrant inclusion as an author. There may be situations where those providing administrative support attempt to have their name included as co-authors of papers for less than honest reasons. An example of misleading authorship is the inclusion of members of an academic department on papers to which they have not made a contribution. This may occasionally be done to help include them in national research audits. This is particularly dishonest because it is an unfair and fraudulent attempt to deceive those making decisions about how public funds will be spent.

Large-scale research projects, whether externally funded or not, may involve a team of researchers from different institutions, both academic and commercial. However, not all project personnel will be involved in every paper that is produced from the research project. Only those people who undertook the research or made other academic contributions to the planning, design, analysis and interpretation of the results reported in a given paper should be included as authors. The inclusion of 'passengers' is dishonest and unfairly credits them with achievements they have not made.

This author uses the term 'hijacker' where an academic takes a dissertation they have neither done themselves nor even supervised and writes it up as a paper. While the process of writing the paper does involve academic work, there is an ethical concern that the academic did this without permission to put their name to research with which they were not involved. An additional member of staff may be brought into the research with the agreement of the student and supervisor to help improve the quality of the paper using the additional academic's expertise in relation to theory and the discussion. It is understood that this type of assistance from the supervisor or any additional member of staff should not occur during the student's dissertation period because the student's sustained independent research is to be assessed. However, where an academic 'hijacks' a student's research without permission, it is a case of dishonest practice. The student and supervisor could have the ability to write the study up as a paper but may not wish to if their careers are taking other directions and they have other more important priorities.

Student research projects are part of their programme of study contributing substantially to their final degree classification. A research project is undertaken under the supervision of an appropriate member of academic staff. As already mentioned, the student should be allowed to make mistakes, receive feedback, learn and improve draft chapters before final submission of the thesis. There may be ideas discussed with the supervisor during the early stages that prevent the project going in an unwise direction. Once the project is submitted and the student's work is assessed, it can then be considered for publication. Indeed, there have been situations where research projects warranting marks of less than 60 per cent have been published where the dissertation lost marks for presentation and discussion parts that were rewritten during the process of producing the manuscript for publication.

One thing that members of academic staff must remember is that the student's primary purpose for being at university is to work towards their degree rather than publish papers for the benefit of the university. Some students may freely choose to attempt to have work published during their programme of study and in rare cases the student may drive this process (an example is Rudkin and O'Donoghue, 2008). The supervisor should always advise the student of their priorities and ensure that their efforts towards receiving the degree classification they are capable of are not compromised by such research efforts. Even where the member of staff may have driven the process of publishing a student's research, it is typical in performance analysis for the student's name to appear first because the student did the research and the paper is a report on that research. The supervisor will benefit from having the student's dissertation, often in electronic form, when reformatting it as a paper. Where a supervisor has had to make an academic contribution that exceeds the total academic contribution of the student, there is a case for the supervisor's name appearing before that of the student in the author list. Research-active members of university departments who have good personal research records do not need to engage in such 'hijacking' activities, which are more typical of those with questionable research records.

One of the most serious ways in which 'passengers' and 'excess baggage' authors are included on research outputs is nepotism. Including the name of a relative or friend is nepotism where the relative or friend has made no academic contribution to the work. There are other forms of nepotism, for example where a research-active member of university staff is invited to give a keynote presentation at an international conference on a topic of their choice, and they choose to cover an area that a relative or friend may have been doing simply to include the relative or friend's name as an author and add to their curriculum vitae (résumé). This practice is totally wrong where no additional work has been done by the relative or friend in preparing the keynote presentation. Previous data collected in part or whole by the relative or friend does not warrant their inclusion as an author where no further work has been done by them. Even when the member of staff who is invited to give a keynote presentation persuades their relative or friend to make some small contribution, the practice is still to be frowned upon as the invitation was made to the member of staff alone. Nepotism is not only unethical but it can also be very embarrassing for the relative or friend when they are fully aware that many in the academic community know what is happening and take a dim view.

There are also examples of unethical behaviour where people who have made a genuine academic contribution to a research paper are excluded from the author list. This could happen where a junior member of staff or a member of staff on a fixed-term contract is excluded from an author list by more senior colleagues and feels that protesting against this may be harmful to their career prospects.

Unacceptable publication practices

Unacceptable publication practices are where researchers publish the same research in more than one research outlet. This is dishonest where journals clearly stipulate that submitted research must be original. Paying delegates at research conferences also expect to be presented with up-to-date original research. There are many pressures to publish the same research in more than one outlet. University press offices may also wish to publicise the results of research studies before the academics responsible have published the studies in their chosen research journal. In any such situations, researchers should resist the temptation. There are legitimate cases where data are reanalysed for different research purposes. Thomas *et al.* (2005: 83) stated that good research practice is to include all analysis of the data set in a single primary publication. However, there are often word limits for journal articles that prohibit the publication of all possible analyses of performance analysis data. Furthermore, there are occasions where authors submit a paper analysing the effect of two or more independent factors on a set of performance indicators and are required by the reviewers to exclude the hypotheses, analysis, results and discussion with respect to one or more of these factors in order to tighten the aims of the paper.

ETHICAL PRINCIPLES

Researchers have a responsibility to conduct their research in a manner that minimises physical, social, psychological and financial harm to research participants, fellow researchers, their university and the wider community. The process of obtaining ethical approval from an organisation should not be seen as the end point of ethical issues within the research project. The researcher must engage in an ongoing process of ensuring that the project continues to be conducted in the manner approved by the organisation. Researchers should also encourage collaborators within research projects to act ethically throughout the full duration of joint research projects. Performance analysis researchers should promote ethical research practices and avoid any practices that would bring the performance analysis discipline or sports science into disrepute.

National and international professional bodies in sport and exercise science often have stated codes of conduct that members are obliged to adhere to. For example, the American College of Sports Medicine (ACSM) has a code of ethics for its members, whether practitioners or academic researchers, as well as a code of ethics for certified and registered professionals. The International Society of Performance Analysis of Sport (ISPAS) does not currently have a code of conduct although it does expect accredited performance analysts to work professionally and responsibly. Performance analysis researchers are typically employed by academic institutions rather than by ISPAS, and responsibility for their conduct lies with themselves and their employers who will have ethics policies and procedures in place.

GENERAL ETHICAL ISSUES

The wellbeing of research participants is of the utmost importance and this can often compromise the pursuit of the truth within research. For example, researching coaching practice may uncover failings in coach education programmes, the publication of which would be embarrassing and damaging for individuals. However, such research would also improve knowledge of pitfalls to avoid in coach education programmes. Researchers often find themselves facing difficult ethical dilemmas and need to develop solutions to these problems. Researchers have to compromise their search for the truth in order to proceed with their research in an ethical manner. This approach recognises that undertaking research in an ethical manner is paramount and avoids the potentially damaging consequences of unethical research. There are many sources of ethical problems, some of which can be anticipated before data collection and some which may not be encountered until data are being gathered and analysed. Where ethical problems are anticipated by researchers or members of research ethics committees, these problems can be addressed prior to data collection commencing.

Voluntary informed consent

Those who participate in research studies should do so of their own free will with a full understanding of what they are participating in, especially their own personal involvement in the study. Potential participants should be provided with a complete and clear explanation of the purpose of the study, any source of funding for the study, what participants are required to do, how many occasions they are involved and the duration of each task they are required to do, any risks of any kind that may be involved and any pain or discomfort that may be experienced. The researcher should explain the extent to which participant anonymity and confidentiality will be preserved. There are circumstances where researchers would be obliged to divulge information about participants for legal and professional reasons, so researchers should not give unrealistic guarantees of anonymity and confidentiality. Where data are being gathered using video or audio recording devices, the participants should be made aware of this. Some research projects may be able to provide an option for data to be gathered without such devices, but in other research projects, such devices may be essential. The participants' willingness to have their words and image recorded is something that the researcher must consider when recruiting participants.

The use of the data within the project as well as any possible uses of the data within future research projects, by the researcher or others, should be explained to potential participants. Where the participants have provided their consent for a particular study, it is unethical to then use their data for further studies where they have not provided their consent.

Any benefits of participation should also be clearly explained to avoid disappointment if participants form an unrealistic expectation due to vague descriptions. The explanations of the purpose of the research, what participants will do and any risks involved should be made as far as possible in language that the participants can understand. The researcher should answer any queries that potential participants have about the research and their involvement in it.

People must not be pressurised into participating in research studies and have the right to withdraw without prejudice from the study at any point if they wish. An example of unethical behaviour can occur where students undertaking university degree programmes are recruited as participants in research studies by their teachers. Students may feel that they should participate in such investigations in order to be awarded good marks in coursework and exams. Even where marks awarded are independent of involvement in the study, the students still might feel pressure to participate in the study. Sometimes, data are collected from students as part of a module that they are undertaking. For example, performance analysis students could be asked to analyse a performance in order to learn about a system, the nature of data gathering in performance analysis or reliability assessment. The author has organised such exercises during practical sessions with his own students. The results are used in the classroom to discuss disagreements in data entered between students to learn about sources of error in performance analysis. These data have never been published and will not be published as the author has never applied for ethical clearance to use such data in research and the data came from trainee users. There would be an ethical problem if the gathering of data was primarily being done for research purposes and there was very little educational value in the exercise.

Another way in which people may be pressurised into participating in research studies is through the use of 'gatekeepers' who may have some influence or power over them. The researchers should always seek voluntary informed consent from the individual participants with the gatekeeper's role only being to identify potential participants. Informed consent means that any potential participant must sign a document that states that they understand the purpose of the study, their own involvement and that they are free to withdraw at any time without prejudice. In research involving covert methods and deception, it is not possible to obtain voluntary informed consent, as will be discussed later in this chapter.

There are some studies where participants cannot provide voluntary informed consent, for example under-18-year olds or people who would not have the ability to understand the explanations provided about the research. In such situations, researchers should obtain informed consent from parents, guardians and teachers. There may also be legal requirements associated with working with under-18-year olds in any capacity. For example, in the UK it is necessary to be cleared to work with under-18-year olds by the Criminal Records Bureau (CRB check). There may be a risk of anxiety and stress caused by research participation for those with learning difficulties, for example. The condition experienced by such participants is also a private matter for themselves, their families and those who care for them. Therefore, researchers who wish to include such participants within their investigations should seek the advice of relevant professionals and take the necessary care if using such participants.

Privacy, confidentiality and anonymity

Privacy and confidentiality are serious matters in research; some countries have legal requirements for the recording and storing of personal information such as the UK's 1988 Data Protection Act. There are research journals that will not publish research papers where participants are identified. There are also professional bodies that have codes of conduct forbidding the identification of research participants. This raises issues for many performance analysis investigations where elite participants are identified and discussed within research papers. This occurs where data are already in the public domain through television, newspaper or internet coverage or where spectators are permitted at match venues.

Questionnaires can be provided anonymously by using identifiers that would not indicate who the respondent is. Anonymity is not possible with interview data, making it very important that any interview recordings and transcripts are stored securely and that participants are not identified directly or indirectly by the way a research study is reported. Therefore, research papers, dissertations and other research reports should avoid stating the role or organisational position that might identify a participant. Researchers often work in collaboration with other researchers and professionals from other institutions that may have regulations or agreements with clients about the confidentiality of information. For example, the player tracking data of the ProZone3® system are only available to the two clubs participating in the match and are not disclosed to third parties without the agreement of both clubs to share the information.

Invasive testing and measurement

Research can be argued to be invasive where participants are required to spend time and effort performing tasks that they would not be doing if they were not participants in the given research study. Therefore, the use of interviews, focus groups and questionnaires can be said to be invasive research methods. This section is concerned with testing and measurement that involve the participants performing activity in laboratory or field settings that is much more disruptive to their routine activity than completing a questionnaire would be. Examples of such methods are:

- phlebotomy;
- taking muscle biopsies;
- taking saliva samples or other human tissue;
- anthropometric measurement and fitness testing;
- biomechanical analysis outside competition situations;
- dietary interventions;
- exercise prescription;

- performing psychological skills training for the purpose of the investigation;
- wearing heart monitors or other devices during training, competition, work or free time that would not normally be worn.

Some of these invasive methods involve a risk of physical or psychological harm, while others involve some discomfort. The researcher should do a risk assessment of any activity to be undertaken by the participants and ensure that any risks involved are explained to potential participants. Where strenuous exercise is involved, it may also be necessary to ensure as far as possible that participants obtain medical clearance. Exercise testing and measurement should be done by appropriately qualified personnel and if any substances are being produced to be ingested by the participants, these should also be prepared by qualified personnel. The general process of obtaining ethical clearance for such research is discussed later in this chapter; however, researchers must use the procedures that operate within their own university or organisation.

Vulnerable populations

Vulnerable populations include under-18-year olds, the elderly, people with learning difficulties and the physically or mentally disabled. As mentioned earlier, researchers must obtain voluntary informed consent from parents, guardian or teachers when investigating under-18-year olds. Where participants are recruited and voluntary informed consent is provided by the parent or guardian, the researcher should not use this to coerce a member of a vulnerable population to participate against their will. There may be occasions where a participant states that they do not wish to be involved in a study or it is apparent that they do not wish to be involved. In such situations, the participant should be allowed to withdraw from the study even if parents, guardians, teachers or carers have given their consent.

Deceptive research

There is a range of deception that can occur within research projects. An example of a non-contentious use of deception was a study of relative age effect (Perham and O'Donoghue, 2009) where a series of interviews were used to investigate the lived experience of field hockey players born in different halves of the junior competition year. The student was concerned that the responses made by participants might be influenced by the purpose of the study and, therefore, the precise purpose of the study was concealed from the participants until the final question, when the student asked what month the participant was born. This was done to maximise the chances of uncovering the truth about the differences in the experience of sport between those born in the first half of the junior competition year and those born in the second half. The general aim of the study, to investigate the lived experience of field hockey players, was explained to the participants, although the authors did not completely obtain full voluntary informed consent. One way around this problem would be to explain the full purpose after the study and seek voluntary informed consent to use the interview data at that point.

Covert methods are used to investigate issues where the pursuit of the truth would be seriously compromised if participants were aware they were being investigated. The researcher gains access to a group of interest without the group providing any consent for the research study to take place. The researcher then investigates the group over a natural cycle of time through a combination of observation and conversations. The research is deceptive because the researcher conceals their true role and acts as a member or associate of the group. This requires a combination of good acting skills and naïve participants. The ethical issues are that the participants do not realise they are participating in a research study and have not given their consent to be studied, the participants' privacy is violated, the researcher is potentially placed in danger and the participants may experience embarrassment or other damaging consequences on publication of the study. As will be mentioned later in this chapter, many university research ethics committees will only approve deceptive research if the knowledge to be discovered by the research is very important and there is no other way of investigating the problem.

Sensitive aspects of behaviour

Sensitive aspects of participant behaviour can be investigated using a variety of methods, including interviews. The ethical issue is that potentially private and possibly illegal behaviour of participants is being investigated. Even where overt methods are being used, the investigation of issues such as drug use or hooliganism requires participant consent, which may be difficult to negotiate. Furthermore, if the research is inductive rather than testing fixed hypotheses, the initial research proposal cannot specify the full detail of the participants' behaviour or how it will be portrayed in any eventual research report. The conclusions drawn from the study may portray an image of the participants that they do not accept.

SPECIFIC ETHICAL ISSUES WITHIN PERFORMANCE ANALYSIS RESEARCH

The performance analysis discipline does not involve the same ethical issues as deceptive socio-cultural research or laboratory-based biochemical and physiology studies. However, there are investigations that combine performance analysis methods with methods from other disciplines and so the relevant ethical issues need be addressed. The most common types of performance analysis investigations are minimally invasive observational studies. The main ethical issues involved in such studies relate to filming, analysis of vulnerable populations, the use of publicly available data, reporting sensitive information from case studies, anonymity of participants and access to performance data.

Filming and audio recording

Filming at sports venues is often prohibited due to contractual agreements between sports governing bodies, clubs and commercial broadcasters. Clubs and governing bodies may also sell video recordings of performances and it is necessary to protect valuable material from piracy. Indeed some professional soccer clubs prohibit any recording of information by spectators. For example, a Cardiff City Football ticket for the 2008–09 season states the terms and conditions that are accepted by the purchaser:

> Number 2
> No person may bring into the ground any equipment which is capable of recording or transmitting audio or visual material or any information or data relating to any match, or the ground. Mobile phones are permitted for personal use only.

> Number 7
> The use of this ticket to enter the ground constitutes acceptance of such rules and regulations and CCFC reserves the right to eject from the ground, any person who fails to comply with them.

Where filming of sports performance is necessary within a research project, the filming must be carried out in an ethical manner, respecting the privacy of individuals and making responsible use of equipment. Before contacting sports organisations about filming, the researcher should investigate the regulations regarding filming at particular venues, competitions and for the sports clubs and governing bodies involved. The principle of voluntary informed consent applies to filming just as it applies to surveying, testing and measurement in research. In writing to a sports club or event organiser to request permission to film, it is necessary to describe the purpose of the investigation so that those granting permission will be fully informed about the purpose of the study and what is being filmed. Assurances must be given about the use of the video footage being restricted to the research project. Filming must be done in a way that minimises risk of damage to the equipment, harm to the researcher, players and other spectators. Equipment must be checked and certified as safe to use at the venue. Electricity cables must be taped down to avoid people accidentally tripping over them and filming must be done from a location approved by the venue. Researchers given permission to film at a venue must cooperate fully with events staff at the venue. Many venues have child protection policies and procedures requiring anyone taking photographs or video recordings to complete forms with details of what is being filmed, the purpose of the filming, the use of the images as well as contact details. Once such a form has been signed and approved by the staff at the venue, a badge or label is provided that must either be displayed on the researcher's clothing or on the camera. It is important to also consider other spectators when recording information. For example, if using an audio recording device to verbally code observed behaviour, this must be done in a way that does not interfere with paying spectators' enjoyment of the match.

Performance analysis support

Studying the effectiveness of performance analysis of sport typically involves providing a performance analysis service to players and teams. The perceptions of players and teams of the feedback given can be studied (Jenkins *et al.*, 2007) as can improvements in actual performance (Jenkins *et al.*, 2007, Martin *et al.*, 2004). These studies can involve the researcher keeping a log of analysis results and how these are used to inform coach decisions. Alternatively, qualitative field notes about the experience of working as an analyst can be produced. There is a vast amount of complex information that can be recorded within each match to match cycle including the following data:

- Performance statistics from the match just played.
- Players available to compete in the match and their fitness levels.
- The quality and style of play of the opposition.
- The detailed aspects of technique or tactics requiring attention that are identified by the coach from match statistics and related video sequences.
- Whether and how these aspects of the game were addressed in short-term training activity and advice to players.
- Reasons for the priorities set by the coach for different areas to be addressed in training.
- Coach and player perceptions of the usefulness of the instructional feedback provided (statistics and/or related video sequences).
- Coach and player reaction to motivational feedback provided.

There are many squads and individual athletes in different sports who recognise the benefits of performance analysis support but who do not have the means to pay for the service. There are ways in which basic performance analysis services can be supplied by volunteers or performance analysis students requiring work experience. One way in which support can be provided is by using university staff or students who wish to study the

effectiveness of performance analysis support. This can be a mutually beneficial arrangement but does give rise to some ethical questions.

First, if the coach and the squad fail to use performance analysis support
in an effective way, the researcher faces a dilemma as to whether to write up
the study portraying the squad's use of the service in a negative way or to
conceal genuine findings of the study. The researcher may even witness some
dishonest practices by the coach where players are intentionally provided
with false information about their performances. There is also incompetent
use of performance analysis where a coach may point to an improvement in
a performance indicator that may be purely down to playing world class
opposition in the previous match. There may be issues in the squad that
affect performance that the researcher wishes to explain in the study. For
example, a drinking culture within the squad may cancel out any advantages of performance analysis and other sports science support. The coach
may object to the publication of such material and may wish to censor any
material being written up as research. The ethical issues here apply to the
researcher and to the coach. The researcher may have worked for a pro-
longed period of time gathering real-world data to write an experience
report about the use of performance analysis. This may have been done at
the expense of other career development opportunities. The coach may have
been involved in a vague agreement to permit such material to be reported.
However, on discovering the detail of the research findings, the coach may
prohibit the publication of the report. This may be viewed as unfair by the
researcher who provided a service on a voluntary basis on an understanding
that material could be used in their research. To avoid such situations, it is
important that at the outset of the collaboration, the researcher, coach and
athletes have a full understanding of the scope and purpose of the study and
the material that the researcher is permitted to use.

Secondly, there is the issue of confidentiality. The performance analysis
system is typically developed and co-owned by the entire squad, including
the performance analyst. There may be types of analysis done that give the
squad an advantage over their opponents. This effective process of analysis
is something that the squad may wish to keep secret but which the researcher
may wish to write up as an example of best practice. There may be ways in
which the researcher can describe the development and operation of the
performance analysis support in vague terms, but this will affect the chances
of the research being published. The potential of such conflicts arising
between the squad's need for confidentiality and the researcher's need to
write a complete and accurate research paper can be reduced by detailed
discussion and agreement of what may be used as research material at the
outset of the collaboration.

Thirdly, there may be failings by the analyst. These may include filming
errors, coding errors, failing to provide feedback in time, favouritism to some
players through providing them with additional feedback material, dishonest
practices related to known limits in reliability and there may be health and

safety issues in the way equipment was used. In providing a performance analysis service, the researcher should not only act ethically as a researcher but should also act professionally and responsibly as a performance analyst.

Using public domain data

There is a wealth of publicly available sports performance data that are used in performance analysis investigations of tennis (Brown and O'Donoghue, 2008b), soccer (Rowlinson and O'Donoghue, 2009), netball (O'Donoghue *et al.*, 2008), athletics (Brown, 2005) and many other sports. Professional players are paid by their clubs or receive appearance money or prize money from the events they compete in. These professional clubs and sports events receive money from broadcasting companies which in turn receive money from the subscribers to the given television channels. In signing for such a club or entering such a tournament, the player has already agreed to participate in matches that may be televised and have thus given their consent for information about their performances to be made public. Therefore, academic researchers do not need to obtain voluntary informed consent to use publicly available television images of performance within research investigations. Indeed, there are investigations that have identified particular athletes (Brown, 2005, Hughes and Franks, 2004d, O'Donoghue, 2004, 2009a).

There are some ethical issues in using publicly available material that should be recognised by performance analysis researchers. Broadcast material is often provided commercially and, therefore, researchers or their universities should subscribe to the television channels that provide the material. The material should not be passed on to other researchers who may not be subscribers to the television channel. The researcher should also avoid presenting results in a manner that could unfairly risk social and financial harm to players, coaches or referees.

Some research journals and professional bodies prohibit the naming of individual participants within papers and other research outputs. Others provide conditions requiring participants to give permission to be identified in the paper. For example, the author guidelines for *Medicine and Science in Sport and Exercise* (2009) contain the following statement within the 'Policy Statement Regarding the Use of Human Subjects and Informed Consent':

> An explanation of the procedures to be taken to ensure the confidentiality of the data and information to be derived from the subject. If subjects are to be identified by name in the manuscript, permission for same should be obtained in the Informed Consent Form or obtained in writing at a later date.

Where journals prohibit the identification of players, researchers whose work identifies players should choose other outlets in which to publish their research.

ETHICS POLICIES AND PROCEDURES

Research ethics committees

Universities, other academic institutions and health authorities have research ethics committees who assist researchers in undertaking research in an ethical manner by developing and operating research ethics policies and procedures. A university will typically have a university research ethics committee as well as research ethics committees for individual departments (or faculties or schools). Research-active staff and students submit research proposals to these committees to seek approval to commence data gathering. Research proposals with major ethical issues will be submitted to the university's research ethics committee while more routine projects will be considered at department level. Where projects are approved at department level, an audit should be maintained and decisions reported to the university's research ethics committee. Departments only need to consider those projects where there are ethical concerns that cannot be dealt with by the supervisor. Where the supervisor is able to recommend ethical approval, it is still necessary for the departmental research ethics committee to receive details of the project so that all such recommendations can be formally approved and recorded. Therefore, all research being done within the university will be recorded as being approved at some level and this information will be maintained by the university's research ethics committee. There are also research proposals that are not approved, appeals processes and opportunities for resubmission. In addition to the processing of research proposals, research ethics committees also have a role in promoting ethical research practices and awareness of ethics policies and procedures. Research ethics committees will produce the necessary documentation to assist staff and students develop the awareness of ethics policies and procedures within the university. This documentation typically includes handbooks, ethical application forms, guidance notes and internet resources.

At departmental level, the composition of the research ethics committee depends on a number of factors including the size of the department, whether dissertations are compulsory within taught programmes and the type of research that is done within the department. In a large multidisciplinary sports studies department of more than 60 academic staff, where more than 400 undergraduate research proposals, additional Master's and research degree proposals have to be considered, the department's research ethics committee will represent all research disciplines within the department. Where there are several members of staff in each discipline, having a discipline expert in university research policy and procedures helps to promote ethical research within the department and makes the process of scrutinising research proposals a reasonably smooth one. In small sports studies teaching teams of fewer than 15 members of academic staff, it may be more efficient to have a single expert in university research ethics policies and procedures, who is part of a departmental research ethics committee within a department that comprises several areas including sports studies. Some university sports studies departments are involved in laboratory-based studies with serious ethical issues, more so than other university sports studies departments. This is also true of other areas within sports studies that have issues to be addressed.

Scrutinising research proposals

Research is undertaken by academic members of staff, postgraduate research students, Master's degree students and undergraduate degree students. All research that is done within a university, whether it involves ethical issues or not, must obtain ethical clearance by the university before participants are recruited and certainly before data collection can commence. The way in which the research of academic staff is scrutinised by research ethics committees varies between universities. There are fewer members of academic staff than students and, therefore, there is a possibility of all staff research being submitted to the departmental research ethics committee. This has the advantage of familiarising all research-active staff with the ethics policies and procedures, which in turn helps those staff when dealing with student research proposals. Some universities may choose to use similar processes to approve staff research and student research. A necessary difference is that instead of staff research proposals initially being considered by a supervisor, they can initially be considered by the appropriate member of the department's research ethics committee. This member of the department's research ethics committee acts as a gatekeeper to the ethics process for staff. Postgraduate research students may also use a gatekeeper such as the chair of the department's research committee.

The purpose of the initial submission of research proposals is to determine whether they need to be considered by the department's research ethics committee, the university's research ethics committee, or if they can be recommended for approval by a supervisor or gatekeeper. Proposals would be sent to the university's research ethics committee if they have serious ethical concerns such as:

- Analysis of blood samples or human tissue.
- Invasive testing or measurement.
- Repetitive testing requiring a greater volume or intensity of exercise than the participants would normally experience.
- Administration of drugs or other substances in dosages that are not normal for the participants or drugs or other substances that are not commercially available.
- Studies that could cause greater physical or psychological stress and anxiety than participants would normally experience. Some interview

studies have the potential to cause stress to participants (Kvale and Brinkmann, 2009: 63).

- Investigating vulnerable populations or participants who are unable to give voluntary informed consent.
- Covert research.
- Investigations involving health service patients, staff or facilities.

Proposed research involving any of these serious ethical concerns can be done at universities, provided the university's research ethics committee is satisfied that such research will be undertaken in an ethical manner, by suitably qualified and experienced researchers, minimising potential risks to the participants, the researcher, the university and wider community. Research proposals with any of these serious ethical concerns must be specified in detail so that the university's research ethics committee will be able to make a decision based on the proposed methods. Where a proposal is vague or incomplete, it will not be possible to make a decision on the research. For this reason, departmental research committees often check such proposals to ensure that only competently written proposals are submitted to the university's research ethics committee.

Different countries have different laws relating to the use of blood and human tissue in research. University research ethics committees will reject any proposed research that potentially involves illegal gathering or storage of blood or human tissue. Any proposal to use deception will only be approved if the research question is very important and the researcher has justified that there is no alternative way of obtaining the necessary data. University research ethics committees will typically encourage researchers to use role playing or other alternatives to deception for investigating the research question. Deception violates the principles of informed consent and may also be illegal if participants' privacy is violated in non-public places. Where ethical approval is granted for covert methods, the university research ethics committee may impose strict conditions that consent is obtained retrospectively. Where health service premises, facilities, patients or staff are involved in research studies, the university will also require researchers to have obtained ethical approval from the appropriate health authority.

Departmental research ethics committees will scrutinise any research proposals with other ethical concerns that do not need to be sent to the university's research ethics committee. These include:

- investigations involving children in school or other public places;
- testing that involves similar volumes, intensities and exercise protocols that would normally be experienced by participants;
- investigations of sensitive aspects of participant behaviour.

Schools often have their own procedures for approving the involvement of school children in research projects. This may cause problems if schools refuse to operate within the procedures used by the university. In such cases, the approval processes operated by the school may be used and may be sufficient to satisfy the ethics committees within universities.

Projects with very minor ethical concerns or projects that follow standard protocols may be recommended for approval by potential supervisors when the proposals are assessed as part of a dissertation module or research methods module. The recommended decision is formally approved at departmental level, which means that the recommendations of members of staff are trusted. Therefore, members of academic staff must be acquainted with ethical issues and policies in their research areas.

One issue that needs to be considered in this process is the level of detail included in the research proposal. There are research projects that cannot be approved if insufficient detail is provided. In cases where a research proposal is vague but the assessor is confident that whatever is done will not involve data gathering or analysis methods that would raise ethical concerns, there is still a way in which the project can be recommended for ethical approval. The assessor can recommend the project for approval under stated conditions that must not be breached during the research project. For example, a vague research proposal might indicate that the project will analyse performance indicators relating to the tactics of winning and losing teams within UEFA European Champions League matches, but there may be little detail of the matches to be used, the performance indicators involved, whether the data is coming from public sources or how the data will be analysed. There is a balance that needs to be achieved between the need to assess the student's ability to write a research proposal, assisting the student in producing a research proposal so that a decision can be made on it, and allowing research projects to commence without undue delay. In the case described, the assessor awards the mark merited by the research proposal based on the assessment criteria for that piece of coursework. This should be a separate issue to the ethical approval process. A research proposal could fail as a piece of research methods coursework and be approved as a potential dissertation. Alternatively, the research proposal could pass but the project described might not be immediately granted ethical approval. The desirable situation for the student and staff is that the research proposal is of a high enough standard to pass as a piece of coursework and be a research project for which ethical approval can be granted. In the case described, the assessor could recommend ethical approval on the condition that the proposed research uses publicly available broadcast footage only.

Where filming by the researcher is indicated in the research proposal but the proposal is vague, the assessor can recommend ethical approval on the condition that letters requesting permission to film are produced and used, the study does not involve any under-18-year old participants and the video material gathered will only be used for the purpose of the investigation.

Challenges for research policies and procedures

There are many occasions where the need for ethical approval can cause difficulties for the normal teaching, learning and assessment activities of university sports studies departments. One such challenge is where research proposals are submitted as part of a level 2 research methods module but where the university also accepts students directly entering the degree programme in level 3 who may have done levels 1 and 2 at other academic institutions where a level 2 research proposal was not required. These direct entry students need to be provided with an opportunity to have a dissertation proposal approved early during their level 3 academic year. These students will need to submit a research proposal as part of the dissertation module that will satisfy the requirements of research ethics committees and not necessarily include all of the theoretical aspects covered within research proposals normally submitted in a level 2 research methods module. Essentially, the purpose of the research proposal in level 2 is primarily to assess learning outcomes for research methods, with the secondary objective of preparing the students to commence a research project in level 3. The students who enter the university directly into level 3 of a degree programme will have already satisfied the prerequisite learning outcomes for research methods using other forms of assessment. Therefore, a more efficient research proposal should be produced by direct entry students for the purposes of determining a dissertation topic, identifying a supervisor and seeking ethical approval.

A related challenge to that posed by direct entry students is where students change their minds about what they wish to do within their research projects between levels 2 and 3 of their degree programmes. Some students may genuinely not be certain about what they wish to do by the end of level 2 of their programme of study but may have to submit a research proposal as part of the assessment of their research methods module. Students should be encouraged to identify their level 3 research area during level 2, but it is also necessary to provide support for those students who wish to investigate a different area to that described in their level 2 research proposals. Often students change their minds for sound academic or career reasons. In such cases, the opportunities provided for direct entry students to submit research proposals at the beginning of level 3 should also be provided to students who submitted a level 2 research proposal but who have decided to do an alternative research project.

A further challenge arises from the conflict between assessing student ability and ensuring research proposals are written satisfactorily to be submitted to research ethics committees. A research question that is not a worthwhile or important research question can potentially waste participants' time if the research project goes ahead. Where there are no ethical concerns with the proposed project other than wasting participants' time within a poorly designed or unimportant research project, assessors experience a dilemma between wanting to assess the student's ability and awarding a fair grade, and wanting to avoid an ill-advised research project being undertaken. Assessing a student's proposal fairly and providing feedback to assist the student's learning is a normal educational process. This allows the student to revise the proposal in line with the feedback given.

Occasionally research ethics committees question the data analysis procedures described in research proposals. Some sports studies departments may be happy for students to gather data in an ethical manner minimising any risks, but making errors when data are analysed. The analysis of data and the reporting of results seldom involve additional risks to participants or the researcher. Where a research ethics committee forces a student to alter the data analysis procedures before the project is approved on ethical grounds, the student is essentially being given advice about data analysis that is not provided to other students. This advice could come from a supervisor who is keen for the research project to commence or from a research ethics committee where detailed feedback is provided. This conflict between the need to assess student performance fairly and the desire not to waste participant's time in poor research is something that needs to be managed by programme committees and research ethics committees.

SUMMARY

Ethics has been covered from two perspectives in this chapter. First, scientific dishonesty has been discussed with specific reference to the types of dishonest practices that could potentially occur in performance analysis investigations. Misleading authorship and plagiarism are wrong because they result in people gaining credit for research that is not their own. Fabrication of data, omission of data, using data with known reliability problems or where the storage of data may have corrupted values is wrong because readers may make decisions based on the results of the study, believing the data are real and reliable. A further source of scientific dishonesty is publishing non-original research in outlets where it is clearly stated that contributions should be original.

The second part of the chapter described the research ethics policies and procedures operated by universities to ensure that research is undertaken in a way that minimises risks to participants, the researchers, the university, collaborating bodies, the wider community and the research area. All research carried out by staff, postgraduate research students and students on taught programmes must be granted ethical approval before participants can be recruited. The main ethical issues in performance analysis research relate to confidentiality, filming, vulnerable populations and reporting results that could cause social, psychological or financial harm to participants.

CHAPTER 6

DEVISING METHODS

INTRODUCTION

Having selected and stated a research question, whether involving performance indicators or more qualitative information, the researcher needs to devise a method to allow the data to be gathered and analysed to answer the research question. Once a tried and tested method has been developed, the user will understand how long it will take to record data from a single match and hence how many matches can feasibly be included in the study. There are some investigations where the use of technology is essential for reliability. There are other investigations that do not warrant the use of computerised performance analysis systems. There may be existing systems that can be used within the study or new systems may have to be developed for the purpose of the study. This is the case whether manual or computerised methods are to be used. System development, pilot studies and reliability testing are covered in this chapter. Then the main study is considered, covering participants and matches as well as data analysis options. Once the methods are designed, the actual data gathering and analysis can be done.

The activities involved in a performance analysis research project are influenced by the purpose of the study, with some projects being purely observational, others involving supplementary interview or questionnaire components and some involving a degree of invasiveness if monitoring equipment is to be worn by the participants or if data collection is to be done in a laboratory setting. The type of data that are to be captured is a design issue that also influences the tasks to be performed during the research project, with some projects using purely notation data, some using purely qualitative data and others using a mixed approach. There are some performance analysis research projects where systems have to be developed and others where existing systems can be used. Therefore, different research projects will have different structures to the results chapter, as described in Chapter 10.

RESEARCH DESIGN AND PLANNING

Once the student has formulated a research question and they know 'what' the project is investigating, the next step is to think about 'how' the project can be done. This involves devising methods in sufficient detail for an ethics committee to be able to make a decision on the project. The following steps are used in the design of a typical manual notation project where data collection is restricted to publicly broadcast performances:

- Consider the purpose of the study, the hypotheses and sketch the eventual format of the results. This involves thinking forwards and backwards at the same time: forwards to the eventual results and backwards to the purpose of the study.
- Identify the information to be derived from the manual notation system.
- If the variables to be used have not already been defined during the process of specifying the research question described in Chapter 4, then they should be defined when devising the methods.
- The raw data about events occurring within performances that need to be recorded in order to produce the information required should be identified.
- Once the student is aware of the data that need to be gathered, a system can be developed to allow these data to be recorded. The system will contain specifically designed forms to allow the necessary data to be recorded as simply and efficiently as possible, in a manner that also eases the process of determining system output information from the raw data.
- The system must then be pilot tested using a sufficient period of a match to ensure that it is usable and efficient and that the data required are able to be observed from the broadcast coverage of the performances. The system may have to be revised several times until the student and supervisor are happy with it.
- It is necessary to demonstrate that the system produces results that are independent of an individual user's perception of behaviour during sports performance. This requires an inter-operator reliability study to be done. The student must train another operator to use the system and then the system must be used independently by the student and the other operator to analyse the same performance. The performance

analysed may be a whole match, a part of a match or indeed it may be necessary to include more than one performance in the data for the reliability test. The two independent observations are compared to gauge the level of reliability of the system. Where the system is unreliable for some or all variables being assessed, it is necessary to either revise the system or provide additional training to the observer and redo the reliability study.

- Once the system has been finalised, the student should consider the main study of the research project. There is a balance between the number of performances that are needed to answer the research question and the number of performances that can be analysed within the time available for the research project. This can be discussed, because after piloting and reliability testing, the student will be aware of how long is required to analyse each performance. In extreme cases, it may be necessary to modify the system again to increase the number of performances that can be included in the study or to make the system cover more areas of performance as the project in its current form may not be considered ambitious enough for the student's programme of study.

- The data are then collected using the manual system, with individual match data processing being done as soon as possible after each match and an electronic data sheet of resulting match information is populated. This data sheet should be backed up periodically and the raw data collection forms should be kept safe for verification purposes.

- The data are then transferred to a statistical analysis software package where the necessary statistical tests are performed. These systems usually generate a great deal of information, so the student must extract those descriptive and inferential statistics required to produce the results chapter. This is discussed in Chapter 8.

- The methods chapter of the dissertation should be written up in the past tense as the eventual report will be describing research that has been done. It is a good idea for the student to make notes during system development and pilot testing on any issues that can be described in the methods chapter as these might be forgotten if the methods chapter is being written up after the data have been collected and analysed.

Projects that combine the development and use of manual notational analysis with other techniques will still have to go through the above stages. The difference is that the proportion of dissertation time to be devoted to the notational analysis and other elements needs to be decided when designing the project. There are additional stages for the non-notational analysis activities that are part of the study. If qualitative techniques are being used, they could be applied before, after or in parallel with the notation analysis part of the study. This will have implications for the scheduling of activities within the project plan. The qualitative techniques themselves involve activities that need to be scheduled within the overall research plan. For example, an interview study would involve the preparation of an interview guide, pilot interviewing, establishing trustworthiness of data, recruiting participants, conducting the interviews as well as transcribing and analysing the interviews. The complementary use of notational analysis and qualitative techniques will hopefully not be done in a disjointed fashion. Therefore, there may be an additional task of integrating the two sets of findings, although this is typically embedded within the qualitative analysis that would be explaining the quantitative results. The integration of quantitative and qualitative aspects continues within the discussion.

If a computerised performance analysis system is being developed, whether using commercial sports analysis packages or not, the stages listed previously are still used except that code templates are developed instead of manual data collection forms. The computerised system still needs to be pilot tested and reliability tested. Where a project uses a pre-existing system, the system development activities are not needed. However, the system must still be reliability tested to ensure that the particular student can use the system objectively. Sometimes the entire purpose of a study is to do a multiple match reliability study of a system. The research process follows the same stages as those already described as there will still be a purpose to the study and results to be produced. The main difference is that the reliability study is not an additional step preceding the main study, but the two are the same.

A mistake made by some students is that they use commercial sports analysis systems when they are not required for the research. These students are under the impression that if they use these systems they will be awarded a higher mark for their dissertation. There are some projects where the use of such systems is warranted, but where students use them inappropriately they can actually lose marks. These systems require the development of code sets as well as the capturing, storing and tagging of video footage of the performances. This takes time and delays the production of data and results and can mean that the student is not able to analyse as many matches as they could have if other methods had been used. Therefore, students should always remind themselves of the purpose of their study and consider the most efficient and reliable way of gathering the necessary data to answer the research question. Any data of any type that is not required to produce the results to answer the research question are not relevant and their gathering and processing are a distraction.

Some research projects evaluate performance analysis support to a squad of athletes (Jenkins *et al.*, 2007, Kingston, 2009, Martin *et al.*, 2004). The activities involved in such a project will be slightly different to those of the traditional notational analysis study. Projects evaluating the effectiveness of performance analysis support involve the following stages:

1. The initial stage should still consider the purpose of the study and the eventual results that can be produced from it. In sketching the intended results format, there may be a section of qualitative results to be

included, which means that the volume of tables and charts for the quantitative results will be lower than in a traditional notational analysis project.

2. The raw data that are needed to produce the information required should be identified.

3. The student needs to have negotiated providing performance analysis support to the squad before any system development effort commences. The squad need to be aware of what is involved and how it will impact on their preparation for matches.

4. The system development process should involve the coach and the evolution of the system should be documented so that this activity can be evaluated later during the research. Once the system is completed, it should be tested for reliability the same as if it was being used in a traditional notational analysis study.

5. Once the final system can be used reliably, the student should consider the main study of the research project. The system will be used in a cycle of activity that may include some or all of the following:

    a) recording of the performance;
    b) analysing the performance;
    c) providing quantitative results to the coach and related video sequences;
    d) recording the coach's decisions about training priorities based on the analysis;
    e) providing video and possibly statistical feedback to the athletes;
    f) monitoring training sessions before the next match; and
    g) recording any tactical or selection decisions made for the next match as a result of the analysis.

6. The squad performance data and the field note observations about the use of the performance analysis support during the match-to-match cycles are analysed and the results are written up.

In planning the project, any activity or task that takes time must be identified and scheduled. The number of performances to be analysed needs to be determined and particular performances from which data can be gathered need to be identified. Periods of the academic year where the student has examinations to do or coursework to be submitted should be located on a timetable for the year so that research project activity can be scaled down at these times but emphasised at others. Some tasks depend on the completion of other tasks and such temporal relationships between project tasks need to be represented within the project plan. Ultimately, each task should be timetabled within the project plan along with key deliverables such as a completed system, reliability results and findings for the main study. The remainder of this chapter describes the detail of activity involved in formulating the methods.

SYSTEM DEVELOPMENT

Hughes and Franks (2004c) recommended having a clear idea of the information that is required from the system to be developed. Therefore, system outputs should be determined before defining the events to be observed and certainly before designing data collection forms. One of the early tasks that the author asks his own students to do is to get a sheet of A4 paper and divide it into eight sections and then try to identify the six to eight results types that will be produced by the project. This requires the student to think about the purpose of the study, the research question, the hypotheses and the types of result that are needed to answer the research question. In a matter of 15 minutes, the student can determine if there will be too few results, too many results, or if the project will produce a concise set of six to eight tables and/or charts. For example, a study of gender and surface effect on Grand Slam singles tennis performance could be designed to produce the following results types:

- A bar chart of two clusters (gender) of four bars (surface) of mean rally duration.
- A table of other timings for the eight different games (inter-serve time, inter-point time, inter-game time when changing ends and inter-game time when not changing ends).
- A bar chart of two clusters (gender) of four bars (surface) of mean number of shots played per second of rally time.
- A table of service outcome statistics for the eight games (percentage of first serves that were in, percentage of points won when the first serve was in and percentage points won when the second serve was in).
- A compound bar graph of two clusters (gender) of four bar-stacks (surface) showing the percentage breakdown of points into different point types.
- A compound bar graph of two clusters (gender) of four bar-stacks (surface) showing the percentage breakdown of net points into different causes of going to the net.
- A table showing the percentage of points won at the net in the eight different games for the different cause of the player going to the net.

This exercise provides a useful guide to be used when developing the methods; the student must think forwards (to the results format) and backwards (to the research question) when devising the methods. When developing the system, the student should question the system outputs. Are they needed to produce the intended results? Are there any other system outputs that are required to produce the results? Are all of the outputs going to be used to answer the research question? Any system outputs that do not address the research question and will not contribute to the results should be removed from the system and any additional system output needed to fully answer the research question should be added to the system.

Having identified the information to be output by the system and reasoned how this information will be used to produce the results, the student should identify the raw data that are required to produce these outputs. The details associated with each event to be recorded should be considered. Hughes and Franks (2004c) listed position on the playing surface, player, action, outcome and time as key data to be recorded for events within field games. The system must be developed to allow efficient gathering and processing of data and production of information. The student needs to consider if tallying event frequencies is sufficient or if a chronological sequence of events is required. Where the latter is used, it is necessary to include a separate step after data gathering to process the chronological event list to produce the information required. Where temporal analysis of the events is necessary, tallies lose vital data about the order of events. The system may have to process the data in stages; for example McCorry *et al.* (1996) used a set of forms to tally raw frequencies of events in rugby union matches and then used a form that summarised the frequencies from four forms used for each team in each half of a game.

Operational definitions

Any variables that have not already been defined need to be defined during system development. This is required for system outputs as well as the data being entered. Chapters 4 and 7 cover the objective definition of variables. It is worth noting that in performance analysis it is not always possible to be completely objective and very often the quantitative analysis eventually done is an analysis of counts and timings of subjective observer judgements. Consider the definitions of technical effectiveness used by Hughes and Probert (2006) shown in Table 6.1. Terms such as 'slight pressure', 'good technique' and so on are undefined and classification of technical effectiveness is done subjectively. A further issue with this rating system is that two different variables (pressure and quality of technique) have been merged in a restrictive way. For example, excellent technique could be performed under no pressure.

Students should define terms as precisely as possible but should not claim to have made unambiguous definitions where they have not done so. It should also be said that assessors should not penalise students for failing to produce precise and unambiguous operational definitions where this is clearly an impossible task.

Development issues for manual notation systems

When developing a manual notation system, we should consider the eventual results to be produced and how the system can provide information to

Table 6.1 Technical effectiveness ratings for soccer (Hughes and Probert, 2006)

Rating	Operational Definition
+3	Excellent technique performed under pressure
+2	Very good technique performed under slight pressure
+1	Good technique performed under no pressure
0	Average, standard technique
-1	Poor technique performed under pressure
-2	Very poor technique performed under slight pressure
-3	Unacceptable technique performed under no pressure

make the production of the results as efficient as possible. Consider a system to analyse serving effectiveness in tennis where we wish to be able to determine the percentage of points won when:

- serving to the deuce and advantage courts;
- serving to the left, middle and right of the target serving courts;
- using kick serves, slice serves and flat serves;
- serving on first and second serve;
- playing different sets;
- any combination of these five factors.

The student needs to design a form that allows data to be analysed according to the above requirements. The use of tallies allows a near immediate inspection of the results and the requirements in this example do not include any temporal patterns, so the student should avoid using a chronological sequence of points if at all possible. Figure 6.1 shows an example of a form that could be used to gather the data for the project in this example. A copy of this form would be used for each set for each player participating in the match under observation. The use of separate forms for each set prevents individual set information being concealed by overall match totals. Overall match totals can be determined by adding the totals in the individual sets. The form contains 72 different tallies that are accumulated during match observation. However, there are areas of the form where information on different serve types, serving to different courts, serving to different areas of the target service court and first or second service can be summed reasonably easily. The easiest way to compute the percentage of points won in different situations is to key the totals in data entry cells into a spreadsheet that has been programmed to calculate the percentages of interest. When one considers that there are five different factors hypothesised to have an influence on the percentage of points won as well as 10 pairs of factors, 10 triplets of factors, five sets of four factors and one set of all five factors that could potentially be considered, the user needs to determine which combinations of factors they are most interested in determining results for. This is essential to focus the research project and prevent the results from becoming unmanageable.

The student could have created a court diagram to allow the location of services to be plotted. This might be useful in some projects, but in the example described here, we only needed to distinguish between the left, middle and right of the service court. The student should always be looking to make the system as simple as possible (Hughes and Franks, 2004c) and should not be entering data that are not going to be used.

Consider the boxing example described by Hughes and Franks (2004d). This system used shorthand symbols to represent different types of action performed by the contestants. However, the results that Hughes and Franks showed for this match were two pie charts showing the distribution of punches made by each contestant and the following four tables:

1. Total punches thrown by each contestant in each round.
2. The frequency (and percentage) of different types of punch thrown by the two contestants.
3. The number of punches thrown by each contestant when holding.
4. The number of one particular type of punch (the jab) thrown by each contestant in each round.

Instead of having to learn a shorthand notation symbol for each punch type, tallies could have been made in a form like the one shown in Figure 6.2. This would allow a more efficient analysis of the recorded data than the system described by Hughes and Franks. The system they described does admittedly support temporal analysis of punch sequences, which is undeniably important in martial arts, but such temporal analysis was not required for the study they described.

Serve Type	Outcome	Deuce			Advantage		
		Left	Middle	Right	Left	Middle	Right
FIRST SERVE							
Slice	Win						
	Lose						
Kick	Win						
	Lose						
Flat	Win						
	Lose						
SECOND SERVE							
Slice	Win						
	Lose						
Kick	Win						
	Lose						
Flat	Win						
	Lose						

Figure 6.1 Manual form for collecting tennis service data

BOXER: *TYSON*					
ROUND	1	2	3	4	5
Holding					
Jab					
Uppercut					
Hook					
Body					
Miss					
Not Holding					
Jab					
Uppercut					
Hook					
Body					
Miss					
BOXER: *BRUNO*					
ROUND					
Holding					
Jab					
Uppercut					
Hook					
Body					
Miss					
Not Holding					
Jab					
Uppercut					
Hook					
Body					
Miss					

Figure 6.2 Tally system for boxing analysis

Development issues for computerised systems

Computerised systems can be developed by those with programming skills; for example the author developed a computerised notational analysis system for tennis using the Modula2 programming language (O'Donoghue and Ingram, 2001). Today, fourth generation languages exist and database packages allow data entry forms to be developed as graphical user interfaces. The main benefit of this is that the chronological sequence of events can be stored without any loss of temporal information and database querying provides a flexible and efficient way of determining cross-tabulated frequencies. The time-consuming process of producing totals from manual event records that are collected in chronological order is an almost instantaneous process when using a computerised system. It is, therefore, recommended that all computerised sports analysis systems should store details of every individual event entered rather than keeping running totals. The amount of storage required for this quantitative information is trivial by the standards of today's computer hard disks and secondary storage devices. For example, the details of all

36,596 tennis points recorded during O'Donoghue and Ingram's (2001) Grand Slam tennis study can be stored in 7.6 Mbytes of disk space, while the timed sequence of up to 2,000 locomotive movement instances for each of 24 soccer players in O'Donoghue's (1998) soccer study can be stored in 1.4 Mbytes of disk space. Even though the temporal patterns were not being used in these studies, the fact that the systems can efficiently provide the necessary outputs means that there are no disadvantages to storing the data as a chronological sequence of events. There are, however, numerous advantages to storing the data in this form, including:

- Reliability studies can identify all disagreements between independent observations of the same performance. If we merely use the total frequencies recorded for events, there is a danger that some errors might be cancelled out by others, giving an impression that the level of agreement is better than it actually is. This is illustrated in Chapter 7.
- There are potential follow-up studies that can be done if the complete chronological list of events is stored. For example, follow-up studies using O'Donoghue and Ingram's (2001) Grand Slam tennis data included investigations of temporal patterns in tennis points (Wright and O'Donoghue, 1999) and score line effects (O'Donoghue, 2001a, 2003, Scully and O'Donoghue, 1999).
- If video recordings of the match become available, the data can be transformed into a format that can be used to demonstrate integrated match data and video systems.

The computerised data collection system for tennis developed by O'Donoghue and Ingram (2001) is used as an example of a computerised system. This system required some initial match identification details to be entered by the operator before the entry of individual point data commenced. The match identification details included:

- the gender of the players;
- the tournament being played;
- the players contesting the match;
- the player who served first in the match section;
- the score (sets, games and points) at the start of the match section.

The gender was necessary to allow the system to determine whether the match was the best of three or five sets and the tournament was necessary to allow the system to determine whether or not a tiebreaker was played in the final set (this is currently the case at the US Open only). The system required the following details of each point to be entered:

- whether the point emanated from a first or second service;
- timing details (inter-serve times, rally-time and inter-point time);

- number of shots played in the rally;
- type of point (ace, double fault, serve winner, return winner, baseline rally or net point);
- outcome of point (winning player and whether point was won with a winner or an opponent error); this included the outcome of attacking the net;
- the cause of a player attacking the net.

The timing details were entered using the function keys F1, F2 and F3. F1 represents a first service, F2 a second service and F3 the end of the point. The operator of the system pressed these keys while watching the point. The use of function keys to time rallies and other timings also allows the system to determine if a point emanated from a first or second serve. The type of point was derived from the classification of points shown in Figure 6.3, which also identifies the various winning and losing outcomes. This was mapped onto the menu structure implemented within the system.

With players approaching the net, the point classification scheme also addressed the cause and outcome of players approaching the net. The causes of players approaching the net were classified into three types. The first of these is where the player attacks the net to pressurise the opponent. This includes following up a good service, following up a good approach shot

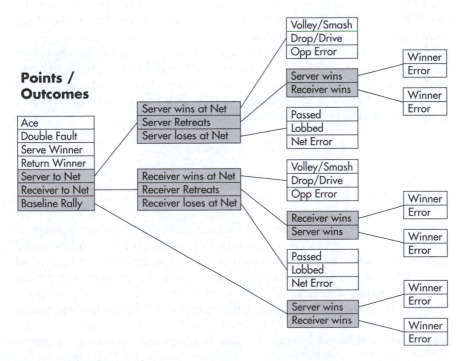

Figure 6.3 Classification of point types in tennis

and 'chipping and charging' to attack an opponent's service. The second cause of approaching the net is where the player is drawn to the net by an opponent's drop shot or short-length shot. The third cause of approaching the net is luck, where neither the player nor the opponent had decided that the player should approach the net. This may occur as the result of the ball being deflected or slowed by the net, or as the result of a miss-hit. The outcomes of approaching the net were classified as follows:

- points won at the net using volleys, overhead shots/smashes, drive winners, drop shots and by opponent errors;
- points lost at the net by being lobbed, passed or making an error at the net;
- retreating from the net, even if the player or opponent eventually returns to the net; only details of the first approach to the net within a point are included.

Computerised systems have the ability to automatically generate values that may be of use in further analyses of the data. As a rule, a computerised system should never require an operator to enter anything that can be computed automatically. O'Donoghue and Ingram's (2001) system automatically updated the score and, where necessary, the serving player. The system also allowed non-scoring points, such as 'lets' to be identified. If a penalty point was awarded, the system allowed the score at the beginning of the next point to be altered accordingly.

The order in which data about an event is entered should map the operator's mental model of the event. This mental model is a sequence of event details that is built up by the observer while observing the event. For example, O'Donoghue and Ingram's (2001) system was developed to record point information in the following sequence:

1. the number of shots played in the point;
2. the point type;
3. the cause of going to the net if the point was a net point;
4. the outcome of the point if it was a baseline or net point.

This matched the mental sequence of information that the observer built up while watching the point. A net point ending with a successful volley would be observed with the operator thinking, 'server to net – following a good approach shot – volley winner'. An early version of the system was almost unusable because it required the user to enter the outcome of the net point and then the cause. Simply changing the order so that the cause of approaching the net was entered before the outcome of the net point made the system straightforward to use because the data entry process matched the operator's mental model of the point.

The operator had to enter the point details during the 20s interval that followed the point. Therefore, the system interface had to be designed in a way to make this as easy as possible for the user. A series of menus were presented, with the user required to type in a numerical code representing the particular option chosen in each case. Using the keyboard was much more efficient than using a graphical user interface. The system updated the score and when the operator saw that the score on the system matched the score called by the umpire at the end of the point, the point data record would be confirmed. If there was any discrepancy here, the operator had to re-enter the point details correctly before the next point started.

The advantages of computerised systems are that they can provide efficient selective searching of large databases of match data. This applies to both scientific investigations as well as in coaching contexts. O'Donoghue and Ingram's (2001) tennis system had a multiple match analysis facility that produced a spreadsheet of variables (columns) by cases (matches) for all 252 matches in their study, which was input into a statistical analysis package. This avoided laborious data transcribing, which would have included errors.

Development issues for systems using commercial packages

There are advantages and disadvantages to using the commercial sports analysis packages that are available today. The main disadvantage is that these systems are designed as general purpose integrated video and match database packages that can be tailored for use with a sport of the user's choice. Such systems cannot be all things to all people; for example, the automatic updating of score line and server in O'Donoghue and Ingram's (2001) special purpose tennis analysis system could not easily be included in a system developed using one of the commercial packages. Another disadvantage is that these systems are used with digital video, which require storage space. A student should not use a commercial video analysis package unless the video component of the system is essential to the research. Capturing video, compressing and then tagging it all require time that delays the production of the main system outputs. This, in turn, can reduce the number of matches that can be analysed within the research project. The projects where the use of such systems is essential include the following:

- Analysis of coach behaviour requires repeated viewing of behaviours in order to classify them (Donnelly and O'Donoghue, 2008).
- Analysis of detailed movements performed by players, as in the Bloomfield Movement Classification (Bloomfield *et al.*, 2004).
- Studying the effectiveness of video analysis (Jenkins *et al.*, 2007).
- Biomechanical analysis of technique where using computerised packages is essential.

- Detailed timing of events in athletics such as touchdown times in hurdles events (Greene et al., 2008).

There are advantages to using commercial video analysis systems, for example such systems may be essential if they are for some of the examples listed above. Another advantage of these systems is that they do not require the student to be able to develop computer software. The use of such packages is also a good skill for the student to have on their curriculum vitae (résumé).

Piloting

No matter whether the system to be used is a manual notation system, a special purpose computerised system or a system implemented using a commercial sports analysis package, the system should still undergo a pilot study to ensure that it can be used for data collection as intended. Pilot studies allow any unforeseen problems to be experienced and rectified before data collection commences. Systems may undergo a series of refinements during pilot testing, with several versions of the system being tested before the final system is agreed (O'Donoghue and Longville, 2004). With word limits imposed on dissertations, it is not possible to outline all of the details of all pilot versions of the system. Indeed, to make the study replicable, only the final version of the system has to be described to the reader. However, it is worth describing the types of refinement made to the system during pilot studies and the student should record this information during system development so that it can be used when writing up the methods chapter.

Description of the final system

The final system must be described in detail, using appendices where necessary. The criteria for recording an event should be outlined and the definition of data variables should be provided. The data collection forms or the interface of the computerised system used should be illustrated, using appendices if necessary. An important point to raise here is that the methods chapter should not read like a user manual. It is not necessary to describe every click of the left or right button of the mouse device when accessing menus or on-screen buttons of a graphical user interface.

In addition to describing the manual or computerised notational analysis system, it is also necessary to describe the observation process and operation of the system during data collection. This is because the operator is part of the system and the success of the system largely depends on the user's ability to observe the necessary detail during sports performances. For example, O'Donoghue and Ingram (2001) described how the trained observer should watch a tennis match to be able to record the cause of a player approaching the net. Tennis spectators habitually watch a player play a shot and then immediately focus on the opponent until the opponent plays a shot. This type of observation risks missing the point at which a player starts to approach the net. O'Donoghue and Ingram (2001) recommended watching a player play a shot and continuing to focus on that player until the ball crosses the net at which point the observer's focus should switch to the opponent. This allows approach shots to be recognised and distinguished from players being drawn to the net during net points.

Reliability evaluation

An inter-operator reliability test is recommended rather than an intra-operator reliability test. This is because an inter-operator reliability test can demonstrate that the system can be operated in a way that produces results that are independent of an individual operator's perception of performance. It is necessary to train another person to use the system and then for the student and the other trained operator to analyse the same match or match section independently. The process of doing a reliability study and evaluating reliability results is covered in Chapter 7.

PARTICIPANTS/MATCHES

Students will read previous dissertations where participants are described before the development of the system to be used to collect data. The fact is that it is important to know how long it will take for a performance to be analysed using the system before decisions are made about the number of participants or matches to be included in the main study. A subsection on matches or participants appearing before the description of the system is not a problem in the final dissertation as long as the decisions about matches were made with the required information being available. A 40 UK credit point undergraduate dissertation module requires 400 effort hours of level 3 work while a 60 UK credit point postgraduate dissertation module requires 600 effort hours of level 4 work. Different projects have different challenges associated with them; for example the main challenge of one dissertation might be the volume of data collection, while the challenge of another dissertation might be a sophisticated data analysis process. Therefore, the number of effort hours to be devoted to data collection may vary from dissertation to dissertation. If a dissertation requires 150 effort hours of data collection and it takes three hours to analyse a single performance, the student will know that 50 performances can be justified as a sufficient volume of matches for the dissertation. However, the student should also consider the purpose of the study and how many performances are needed to satisfy the requirements of any statistical tests to be used. For example, if the student was comparing the work rates of netball players from seven different positions and a Kruskal Wallis H test was to be used, the student would require a minimum of five player performances for each position. The Kruskal Wallis H test and other statistical procedures will be covered in Chapter 8; for now, all we need to be aware of is that we would need a minimum of 35 performances to be analysed in this netball work rate example (seven groups of five performances). Therefore, if the student is tempted to represent each player by an average of five performances, they need to be aware that this would require 175 performances to be analysed, taking 525 hours. This would be too much even for an MSc dissertation. The student can either restrict the number of positions being compared, change the analysis system so that it takes less than three hours to analyse a performance, or use less performances per player. Once the final decision has been made about the number of performances required, the student should ensure that it is feasible to gather data from that number of performances. Is it possible to obtain the data from broadcast coverage? Is it possible to travel to and film the required number of performances? When do the matches take place and can data gathering be scheduled so the correct number of performances can be analysed?

The write up of the methods chapter should contain a section on matches and/or participants. Information should be provided about the performances analysed, specifying the number of teams or players, the number of performances that were analysed for each team or player and the gender, level and age profile of the players who participated in the study. There may be specific criteria for including performances within the study that need to be articulated. For example, O'Donoghue and Ingram (2001) only used sections of tennis matches that were shown live because highlights coverage tended to contain a disproportionately high number of service breaks. The different samples used in the investigation also need to be explained. There are independent samples (sets of completely different matches) and related samples (different conditions within a match or different conditions experienced by performers across different matches).

DATA PROCESSING ANALYSIS

Previously in this chapter, a system was described for gathering data on the distribution of services between the left, middle and right thirds of the target service court under different situations. Once these data have been collected, they are processed in order to produce the results. In the current example, we consider how to compare the distribution of services between men's and women's singles when serving to the deuce and advantage courts on first and second serve. The first question that needs to be considered is the unit of analysis. Four alternatives in this situation are:

1. Gender could be the unit of analysis (with all men's matches being pooled together into a single set of frequency distributions for each condition of interest and the same with the women's matches).
2. Individual player performances (that is both player performances within each match are considered separately).
3. The match is the unit of analysis (with the two players' data being pooled together).
4. Individual players (merging data from different matches into a single set of variables for the player).

In the first alternative, there will be four sets of three frequencies for men's and women's singles; a set of frequencies for serving to the deuce court and to the advantage court on first serve and on second serve. Therefore, the area of the service court where the ball was served to is treated as a nominal variable and a chi-square test of independence can be used to compare a frequency profile of service court areas (left, middle and right) between men's and women's singles. Therefore, a series of four chi-squares tests of independence would be used to determine if the service distribution is influenced by gender or not. A criticism of this type of analysis is that long matches will contribute more to the data than short matches. While there were more points played in such matches, the purpose of the investigation is not to retrospectively analyse individual matches but to use the data in a comparison of men's and women's singles. Therefore, the student should avoid analysis techniques that weight some matches higher than others. This type of analysis is typically used where it has not been possible to gather data from enough matches to reduce those matches to individual performance records and still have enough matches to apply inferential tests on numerical scale variables. This is not necessarily due to poor student effort, but could be because the data gathering process is so time-consuming for each match that the use of the chi-square test of independence violates fewer assumptions of the test than if any other test was being used.

The second way of analysing the data involves initially processing the data so that there are player performance records of 12 values. These 12 values are four sets of percentages of serves played to the left, middle and right of the target service court. There is one set of percentages for each condition of interest (first or second serve to the deuce or advantage court). These percentage variables should be tested to determine if they satisfy the assumptions of parametric statistical tests and then the appropriate parametric or non-parametric tests should be applied. Chapter 8 will cover the detail of these tests, but we will mention here which test would be used in the current situation. If parametric procedures can be used, a series of independent t-tests can be used to compare the percentage of serves played to the three different areas of the service court between men's and women's singles for the four conditions of interest. This gives a total of 12 different independent t-tests. Alternatively, a 2 x 2 x 2 (gender x target service court x service) ANOVA test could test the services to the three different areas. This allows interactions between gender, service and target service court to be investigated. Although the purpose of the study is to compare serving direction between men's and women's singles matches, some might wish to do this by determining if there are significant differences between the areas served to for one gender and not the other. This could involve using a repeated measures ANOVA test, but the student would still need to include gender as an independent factor of the ANOVA in order to compare this main effect of interest. If non-parametric techniques need to be used then a four sets of three Mann Whitney U test could be used to compare corresponding service percentages between men's and women's singles matches. Separating winning and losing player serving performances within games doubles the number of sets of data being included in the analysis and represents the full variability between performances.

The third alternative uses the same statistical procedures as the second except that there is half the number of data values because the winning and losing players' performances in each match are pooled together. The reason for doing this is that low frequencies might be used in determining the percentages if player performances within matches are separated. For example a 6–0, 6–0 women's singles match could have as few as 48 points with an average of six points for each of the four conditions for each of the two players involved. Therefore, the difference of one serve to an area changes the percentage of serves to that area by 16.7 per cent.

The fourth alternative is even better at overcoming the problem of low frequencies as the unit of analysis is the player and the player's record is based on performances in multiple matches. Once each player's 12 percentage values (2 services x 2 target service courts x 3 areas) are determined, the same statistical procedures can be used as described for the second and third alternatives. An advantage of this technique is that if a player is involved in more matches within the data set than other players are, the player will still be considered as a single unit within the data analysis.

DESCRIBING QUALITATIVE METHODS

Chapter 9 is specifically on the use of qualitative research in performance analysis of sport, so this chapter will briefly outline issues in designing qualitative elements of a research study. Qualitative methods may form part or all of the methods used in a research project. There are different types of qualitative research that are distinguished by their theoretical positions (Flick, 2007: 11) and the researcher should first justify the role of the particular type of qualitative method to be used within the project. Sampling is important in qualitative research just as it is in quantitative research. The researcher must determine the participants (as individuals or groups) to be included in the qualitative part of the study, the setting in which this part of the study will occur and negotiate access to the setting and the participants. The analysis of qualitative data, the codes and themes involved, any software packages used and how discrepant cases are dealt with should be recorded during the project. The description of qualitative methods depends on the audience; for a student research project that is being assessed, it is very important to describe the techniques used in the gathering and analysis of data. However, where research is being written up for a research journal, the audience will be very familiar with qualitative research and may, therefore, not be so interested in words being expended on descriptions of methods. This is a point of debate and varying practice, with some journals insisting on the data and methods of analysis being transparent so the readers can decide if they agree with the conclusions drawn within the research.

DOING THE RESEARCH

There are two parts to doing a study: designing the study and doing the study. Once the methods have been devised, there is still more to the research process than simply doing the research as intended. There are three key aspects of doing the research that are considered here: management, ethics and recording.

The chosen methods should be worked into a project plan that schedules the activity involved in the project, identifying tasks and deliverables associated with them. The research plan acts as a vehicle for project management and control, which helps the supervisor and the student discuss progress during meetings. The student should always prepare for supervision meetings by having work done as well as a list of questions or areas for discussion with the supervisor. A further piece of advice is that where draft chapters are being read by the supervisor, these should be sent in advance of meetings so that meetings can be used to provide feedback on the drafts face-to-face.

Once the project has received approval, the student should ensure that the research is conducted in the way that was approved. A signed research ethics form is not just a form to include as an appendix in the eventual dissertation. It is an agreement between the student and the university about how the research can be conducted. It is a disciplinary offence to undertake research outside the terms agreed with the research ethics committee. Where the student needs to use alternative methods or participants for any part of the research, it will be necessary to seek separate ethical approval for such research activity.

During data collection and analysis, the student should keep a log book of activity to record any problems or issues that may be interesting to write up in the eventual dissertation. This is especially true when using qualitative methods where doing the research is more than merely following the research design (Flick, 2007: 39).

SUMMARY

Formulating methods in performance analysis often requires the development of a manual or computerised system to gather performance data. This often introduces the need for a two-part project. First, a system has to be developed and tested, and secondly the system has to be used in the main study of the research project. The system should be specifically developed to be used efficiently and effectively within the research study. This has implications for the data that are recorded using the system, how the data are stored and methods of data analysis. The development of the system requires pilot work, operator training and reliability testing. There may be several pilot versions of the system that are tried before the final system can be agreed. The ease or difficulty of using the system to analyse a performance determines the number of performances that can be analysed within the research project under the time constraints that apply. Where mixed methods are used, the particular methods should be justified and operate in a complementary manner rather than leading to a disjointed research report.

CHAPTER 7

MEASUREMENT ISSUES IN PERFORMANCE ANALYSIS

INTRODUCTION

In Chapter 8, analysis of quantitative data using statistical procedures will be covered. In a scientific investigation, the correct use of the most appropriate descriptive and inferential statistical techniques allows data to be summarised effectively and conclusions to be drawn. However, in any scientific investigation, the analysis is meaningless if the raw data that have been gathered are invalid or inaccurate. In many commercial, industrial and legal scenarios when management information or other evidence is being considered, there are often important questions to be addressed about the sources of evidence. Are the data up to date? What are the definitions of the variables being used? Are the data accurately measured? Consider the example of the university league tables published by the national press in the UK. These have columns such as teaching quality, entry standard of students, research quality, student to staff ratio, library resources and IT resources. When these league tables are published they are often analysed by the senior management of universities. However, these senior officers will not blindly accept the position in the league table quoted for their university. The initial response is usually to question the data reported in the league tables. What is meant by teaching quality? How is it measured? What raw data are used? Are the data valid, reliable and up-to-date? What is meant by student to staff ratio? Are part-time staff included? Are part-time students included? In performance analysis of sport, there are similar measurement issues on which the acceptance of the whole study depends. These measurement issues include validity, objectivity and reliability. Therefore, this chapter precedes the Chapter 8 on quantitative data analysis in recognition that the quality of measurement is of critical importance to a scientific study.

VALIDITY

The word 'validity' can refer to the validity of the whole study or it can refer to the validity of individual variables. When referring to the validity of a whole study, there is ecological validity, catalytic validity, internal validity and external validity, which have been described in Chapter 2. The validity of a variable used in a research study depends on its relevance and its reliability (Morrow Jr et al., 2005: 82). The relevance of a variable is the degree to which the variable represents an important concept being measured. For example, is aerobic endurance an important concept to the area being studied and is VO_2 max a valid measure of aerobic endurance? Is flexibility an important concept to the research area and is the sit and reach test a valid measure of flexibility?

The reliability of a variable in performance analysis is the consistency with which the measurement procedure for the variable can be used by independent operators to measure the same performances. A variable that is not measured reliably cannot be valid, no matter how relevant the variable is to understanding sports performance.

Norm referenced validity

Morrow Jr et al. (2005: 80–125) and Thomas and Nelson (1996: 214–19) classified two broad types of validity: norm referenced validity and domain referenced validity. Norm referenced validity exists where a measured variable can be used to compare a player performance to norms for the relevant population of players. There are four categories of norm referenced validity.

1. *Logical validity or face validity* – is where the variable is valid by definition. This is often the case with performance variables such as 10km running time. There are many outcome indicators in sports performance that have logical validity as they are the score-related variables that the performers seek to maximise, minimise or optimise. The time required to complete a running, cycling, walking or swimming event is a logically valid performance indicator that performers seek to minimise. The distance that a field athlete jumps or throws an object is a logically valid performance indicator that the performer seeks to maximise. The angle of release of a javelin throw is a valid indicator of javelin throwing performance that must be an optimal angle that maximises the distance the javelin is thrown.

2. *Content validity* – is the extent to which the variable (or set of variables) covers different components of the concept of interest. Does a questionnaire

about worry cover all of the areas of worry? Does a test for referees cover all of the situations they will face in a game? In performance analysis investigations, the dependent variables of interest are often a set of performance indicators chosen to cover the broad aspect of sports performance that is of interest to the study. This broad aspect could be strategy, technique, technical effectiveness, work-rate or decision making. In analysing technique, there are many biomechanical indicators of technique including joint and angular displacements, velocities and accelerations as well as kinetic variables. The chosen biomechanical indicators will together have content validity if they cover all relevant details of the technique. A performance profile of technical effectiveness in a team game has content validity if it is composed of technical effectiveness variables for the key skills of that sport.

3. *Criterion validity* – is where the variable is validated against some gold standard measurement that has been accepted as a measure of the concept of interest. The reasons why the gold standard measurement itself cannot be used include the possibility that the gold standard is a very time-consuming measure to apply or involves the use of very expensive equipment or consumable resources. Thomas *et al.* (2005: 194–6) described two main contexts of criterion validity: concurrent and predictive. In concurrent validity, the measurement is correlated against some criteria administered to the same participants within the same study (concurrently). One example of this was the estimation of distance covered by soccer players used in a study by Martin *et al.* (1996). Pre-match speed measures were made and used as velocity multipliers, with the time recorded for different locomotive movements. The product of time and velocity gave an estimate of the distance covered. A more detailed and time-consuming time-motion analysis system was used to analyse the same video recordings of player performances with player locations being entered on an image of the playing surface. The estimates based on velocity multipliers were compared with those derived from entering the path travelled by the players. Predictive validity involves correlating the variable against some gold standard variable and determining a predictive model for the gold standard variable in terms of the variable being validated. Cross-validation is a type of predictive validity where a predictive model is determined based on a subset of the sample of participants and then tested using the remainder of the sample. The test of the predictive model involves making the gold standard measurement for each subject in the remainder of the sample and comparing the actual value with the predicted value using the model based on the variable being validated against it.

4. *Construct validity* – is the validity of some construct used to represent a property that is not directly observable. Construct validity is particularly important in sport and exercise psychology where areas such as anxiety, mood and confidence are measured using questionnaire instruments that compute overall scores for these areas as well as sub-dimensions of them. In performance analysis of sport, the best examples of where construct validity

may be needed are in the evaluation of psychological aspects of performance and in the evaluation of decision making, tactics and strategy. The strategy devised before a match and the moment-to-moment tactical decisions that are made during competition cannot be seen but may be inferred from observable behaviour. Correlation techniques (similar to those used in concurrent validity) can be used to compare the constructs with counts of behaviours one would associate with the construct. For example, if the profile of mood states were used to retrospectively gauge anger during a competition, it might be validated by examining its correlation with behaviours and body language use that would be associated with anger. The degree to which a construct distinguishes between different groups it would be expected to distinguish between (the known group difference method) is also used to evaluate construct validity. In performance analysis, valid outcome indicators would be expected to be different for winning and losing players. Measures of tactics and strategy would be expected to distinguish between athletes and teams who expert opinion would classify as adopting different tactics and strategy.

Criterion referenced validity

In addition to norm referenced validity, there is criterion referenced validity where the measure should accurately indicate whether the necessary level of proficiency has been reached. Decision accuracy is a common type of criterion referenced validity in performance analysis. An example of decision accuracy is the scoring of amateur boxing using the computerised system introduced by IABA (International Amateur Boxing Association) after the 1988 Olympic Games. The system is operated by five judges who use a red button and a blue button to record the punches that are deemed to satisfy the criteria for scoring punches by the boxer in red and the boxer in blue respectively. Where three or more judges press a button of the same colour within a second, a point is awarded to the boxer wearing that colour. The score that is output by this system typically under-estimates the actual number of scoring punches made by each boxer (Coalter et al., 1999). However, as long as the score indicates a win to the boxer who made the most scoring punches, then the system has decision accuracy validity.

Processes of determining valid performance indicators

The dependent variables used in performance analysis investigations are often referred to as 'performance indicators', with some being referred to as 'key performance indicators'. Some students mistakenly refer to the raw performance data that is collected as performance indicators. When a point in a game of tennis is observed, it can be classified as (for example) an ace, a double fault, a serve winner, a return winner, a net point or a baseline rally (O'Donoghue and Ingram, 2001). However, the nominal variable 'point type' used to classify each point is not a performance indicator. The total number of aces served is not a performance indicator because some matches contain more points than others and will have a higher number of aces simply because of the increased number of service points. The percentage of service points where a player serves an ace is a possible performance indicator.

A performance indicator must represent some relevant and important aspect of sports performance in order to be valid. Identifying the valid performance indicators to use in a research project depends on a number of factors that are explained in this chapter. The validity of the performance indicators can be determined through expert coach opinion, review of coaching and performance analysis literature related to the sport of interest, relation to key outcome indicators or discrimination between performers of different levels. In undergraduate performance analysis research projects, there is not sufficient time to quantitatively investigate the validity of performance indicators, unless the whole purpose of the dissertation is to evaluate their validity. Therefore, undergraduate research projects typically select and justify the performance indicators used based on surveying coaches and performance analysis literature or by undertaking preliminary qualitative research to elicit performance indicators from expert coach opinion using a focus group or individual interview. When using performance analysis literature, students will often find that there are no standard performance indicators used in previous published research. For example, when one considers elite tennis strategy, Hughes and Clarke (1995) and O'Donoghue and Ingram (2001) used different variables. Hughes and Clarke (1995) used a combination of rally times, player positioning and shot placement as indicators of strategy. O'Donoghue and Ingram (2001) used rally times and the percentage of points where players attacked the net to characterise strategy. The student should consider which variables are most important to their research, the feasibility of possible methods for collecting the raw data required and the reliability of possible systems and methods that could be used to record the necessary data.

When using coaching literature, whether coaching science research sources or more practical texts and professional coaching resources, the student should consider the aspects of the sport being covered. In non-scientific sources, definitions may be vague and broad areas of technique, tactics, decision making or physical aspects may be written about without identifying any operationalised variables. Therefore, students should use such coaching literature to first identify broad areas of importance within the scope of their research question before considering how these areas can be represented by observable actions that can be counted, timed or assessed. If assessing the effectiveness or quality of an action, it is necessary to consider the number of 'grades' to be used and criteria to be associated with each. Morrow Jr et al. (2005: 138–41) provided examples of guidelines that can be used when setting grades.

Another way of determining performance indicators is to elicit important areas of performance from expert coaches. This can be done during exploratory interviews with individual coaches or using a focus group. The process of turning the identified areas into variables to be analysed within the investigation is similar to when the areas are identified using non-scientific literature. An example of this was an early study of rugby World Cup performance (McCorry *et al.*, 1996) where a rugby expert was initially interviewed about areas of the game that were important to concentrate on when describing the performances in international rugby. This interview was interspersed with periods of watching a video recording of a rugby match, allowing the expert to explain and identify behaviours that were of the greatest importance.

Quantitative methods have also been used to establish the validity of variables used in performance analysis. The process of establishing validity in this way often amounts to gathering the volume of data that a student would be expected to do during an undergraduate research project. Therefore, establishing validity in this way is rarely used as part of an undergraduate student project that has a wider purpose of describing the chosen area of sports performance. It is possible that an undergraduate research project could have the sole purpose of validating a set of performance indicators. At Master's and PhD level, such a validation study could be one of a series of studies that make up the overall research (Choi, 2008). There are different ways in which quantitative methods can be used to establish the validity of performance variables. These include neural networks (Choi *et al.*, 2006b), multiple regression (Choi *et al.*, 2006b), correlation analysis (O'Donoghue, 2002), binary logistic regression, discriminant function analysis and principal components analysis (O'Donoghue, 2008a).

Multiple regression techniques identify the relative contribution of each process indicator in predicting the chosen outcome indicator (Choi *et al.*, 2006b). Choi *et al.* (2006b) found multiple regression to be a more successful predictor of outcome indicators in elite tennis than artificial neural networks. Artificial neural network techniques are also more complex, difficult to use and describe in methods sections of research reports.

Known group difference is a way of establishing the validity of process indicators that can be done using inferential statistical tests. If candidate process indicators are claimed to distinguish between winning and losing performers within matches, statistical tests can be used to confirm or refute this. Similarly, successful and unsuccessful performers can be identified based on finishing position within tournaments and process indicators can be compared between them.

Some valid process indicators are not expected to have an association with match outcome. For example, in tennis there are players who adopt a net strategy in all parts of the World rankings. Similarly, there are players who adopt a baseline strategy in all parts of the World rankings. It is important in practice to understand whether an opponent plays using a net or a baseline strategy. The fact that the percentage of points where a player attacks the net may not be associated with the percentage of points won in a match does not mean that this process indicator is invalid. Similarly, there will be soccer teams that adopt a slow build-up style of play at all levels of the sport and there will be soccer teams that adopt a more direct style of play at all levels of the sport. It is important in practice for soccer squads to have an understanding of the style of play of their opponents even though process indicators representing playing style may not be associated with match outcome.

Statistical techniques for establishing criterion validity and techniques for establishing known group difference often produce sets of process indicators that are not entirely independent (Choi *et al.*, 2006b). Therefore, a more efficient analysis of the given sport can be undertaken if a more concise set of independent process indicators can be identified. Principal components analysis is a data reduction technique that allows a smaller set of principal components to be identified that are uncorrelated variables representing different dimensions in the data. O'Donoghue (2008a) proposed a way in which principal components analysis could be used to determine a set of performance indicators in tennis that represented independent aspects of performance in the sport.

The set of chosen performance indicators should be concise enough to support effective communication but should also have content validity covering all relevant aspects of the area of performance of interest. The performance indicators chosen dictate the action variables that will be used during data gathering. However, an increased number of performance indicators does not necessarily mean that there will be an increased volume of data entry. Consider the POWER system (O'Donoghue *et al.*, 2005a) where operators use two function keys to record when each period of 'work' and 'rest' commences. Originally, this system reported the frequency, mean duration and percentage observation time for 'work' and 'rest'. The enhanced system described by O'Donoghue *et al.* (2005a) included outputs for the frequency of 'work' periods of seven different duration ranges, the frequency of 'rest' periods of eight different duration ranges and 72 frequency variables for each combination of 'work' period duration and following 'rest' period duration. These additional outputs did not require any additional data entry activity by the operators.

DO PRECISE OPERATIONAL DEFINITIONS GUARANTEE GOOD RELIABILITY?

It is essential for system operators and the eventual consumers of the information generated by performance analysis that there is a shared understanding of the variables used. Therefore, there is a view that these variables should be defined with a level of precision that makes their meaning unambiguous (Williams, 2009). The operational definitions used in performance analysis do not have the detail seen in legal contracts, mobile phone contracts or gym membership contracts, and readers are encouraged to examine such contracts to gain an appreciation of how difficult it is to achieve legally binding precision. When it is essential to understand the requirements for software-intensive systems, rigorous formal specification techniques are used based on set theory, mapping functions and temporal logic (O'Donoghue and Murphy, 1996). It may not be wise for operational definitions in performance analysis systems to achieve such levels of unambiguous detail. Performance analysis is often a real-time observational task where there is not sufficient time to consider the level of detail one would see in a legal contract, which is inspected carefully over a much longer timescale before determining if it has been conformed to or breached. Furthermore, such a level of detail would give students and other researchers a serious problem in describing their methods within the required word limits.

O'Donoghue (2007b) provided evidence from two research studies that challenges the traditional view that operational definitions are essential in performance analysis research. One study was an investigation of international rugby union performance (Armitage, 2006) and the other was an analysis of defensive styles used in international netball (Williams and O'Donoghue, 2006). Armitage (2006) produced detailed definitions of named states and transition events in rugby union. The terms defined included 'possession gained during set play', 'possession gained during loose play', 'gaining territory by going over', 'gaining territory by going through' and 'gaining territory by going around' the opposition. The operational definitions came to eight double-sided pages of Armitage's (2006) thesis, with terms introduced during the operational definitions, such as 'gain line', also being defined. Armitage and his supervisor (the author of this book) participated in an inter-operator reliability study. First, the operational definitions were read, discussed and agreed before the final of the 2003 Rugby World Cup was analysed independently by the two observers. The inter-operator reliability study revealed serious limitations in reliability despite the efforts made prior to the inter-operator agreement study. The two operators discussed the reasons why they disagreed so much and what each of them was counting for each type of event. It was concluded that it would have been useful to discuss the operational definitions while viewing example video sequences so the operational definitions would then be considered in terms of the analysis task; an observational task. This would have facilitated learning by both operators of the types of observed behaviours that would be counted for each class of event and for what reasons. One problem with using video observation in this way is that it creates an issue for the replicability of the study as readers of the final report will only see the operational definitions without seeing the example video sequences of the defined behaviours.

The second example used by O'Donoghue (2007b) to challenge the traditional view that operational definitions are essential in performance analysis research was a completely opposite situation to Armitage's (2006) study. Williams and O'Donoghue (2006) did not use any operational definitions of a key independent variable and yet achieved a 100 per cent interoperator agreement for that independent variable. The independent variable was the style of the defence used during opposition possessions in netball. The definitions provided for the four different styles of defence were as follows:

1. Zone defence – where all players concerned in the area of play analysed are marking the space.
2. Man-to-man defence – where all players concerned in the area of play analysed are marking a player
3. Part man-to-man/part zone defence – where some players concerned in the area of play analysed are marking the space and some players concerned are marking a player.
4. Other – where the defence could not be classified in to one of the man-to-man or zonal strategies.

These definitions fail to precisely define the point at which a pattern of seven netball players playing against an opposing pattern of seven netball players changes from one type of defence to another. This is an example of a valid aspect of sports performance that cannot be described precisely or practically in words. One reason for the high level of agreement obtained was that both operators were experienced netball players, coaches and qualified umpires who were very familiar with the terms used and the types of defensive play that counted for each.

Ultimately, any performance analysis technique that involves human operation will involve an element of subjective classification of behaviour. Where fully automated systems such as Hawkeye (Hawkeye Innovations, Winchester, UK) are used in performance analysis research, it is possible to use precise operational definitions for variables. There are examples of situations in sport that are easier for human operators to classify than others. For example, a tennis ball landing in court or out of court is usually straightforward to observe. Furthermore, the observer may decide to use the outcome of the point decided by the line judges and umpire officiating the match. Other aspects of behaviour such as quality of technique, type of technique and locomotive movement classification involve subjective classification by a human observer. Therefore, the timings and frequencies recorded during match observation are often quantitative counts of subjective judgements.

OBJECTIVITY AND RELIABILITY

Reliability of measurement is important in all disciplines of sport and exercise science. In performance analysis of sport, reliability of measurement is a common feature of work in both notational analysis and biomechanics (Bartlett, 2001). Objectivity and reliability are related concepts in performance analysis of sport. A performance indicator is objective if its value is independent of the opinion of individual operators.

The first step in achieving objectivity is to use an objective measurement procedure that does not rely on the subjective opinion of an individual observer. As has already been mentioned, this is often not possible in performance analysis of sport where human observers form part of the measurement system. Objectivity requires the use of definitions for all action variables that will be used in the calculation of a performance indicator. For example, if we have a performance indicator that is the percentage of points where a tennis player goes to the net, we must define what counts as a point and we must define what counts as a net point. O'Donoghue and Ingram (2001) excluded lets (points that had to be replayed) from their study of Grand Slam singles tennis and defined a net point as a point where a player crossed the service line (the line at the back of the service boxes) and there were still one or more shots to be played in the point.

A reliability test seeks to evaluate the consistency with which a method can be used. In other disciplines, such as exercise physiology, test-retest reliability tests are done using the same participants who perform a test more than once. If consistent results are produced for repeated application of the test then the test can be deemed to be reliable. Sports performance variables are different to anthropometric and fitness test variables in that sports performance does not take place under controlled conditions. This allows test-retest studies to be used to investigate the reliability of anthropometric measurement methods and fitness tests. Performance in many sports is unstable and performance indicator values are influenced by many factors including:

- venue;
- players available for selection;
- quality of the opposition;
- importance of the match;
- weather conditions;
- tactical decisions made by coaches and players.

The main factor responsible for variability in sports performance is the quality of the opposition (McGarry and Franks, 1994). One study of variability in sports performance found that David Beckham's percentage match time spent performing high intensity activity in 11 different English FA

Premier League soccer matches had a greater variability than that observed between 115 different English FA Premier League midfielders (O'Donoghue, 2004). Therefore, test-retest reliability studies are not used to evaluate the reliability of action variables, performance indicators or methods of gathering such data. Test-retest studies can be used in investigations of stability of performance, but this is a different issue to the reliability with which a performance can be analysed. Indeed, to be able to study match-to-match variability in player performance, it is essential that the method use to analyse match performances is reliable.

Reliability studies in performance analysis of sport use independent observations of the same performance. This can be done live during a competition or through post-match analysis of the competition. There are two types of reliability study that have been used in performance analysis research: intra-operator agreement studies and inter-operator agreement studies. Intra-operator agreement studies are often used in undergraduate student research projects where it is not feasible to train another operator to use the system developed for the project. However, intra-operator agreement studies are of limited value as they fail to demonstrate the objectivity of the system. A good level of intra-operator agreement in a time-motion analysis system will simply indicate that the particular operator can consistently classify locomotive movement into the defined movement classes. The intra-operator agreement study does not answer the question of whether anybody else using the system to analyse the same performance would obtain similar results. Therefore, inter-operator agreement studies are preferable. Intra-operator agreement studies can be useful for the student during pilot testing of the system.

Inter-operator agreement studies allow the objectivity of systems to be established. They demonstrate that systems can be used consistently, recording data that are independent of individual operator perceptions. Although inter-operator agreement studies are more difficult to organise than intra-operator agreement studies, it is recommended that performance analysis project students use inter-operator agreement tests. One way of doing this is for a pair of students to operate both students' systems during the inter-operator agreement tests. This requires the systems to be described and action variables to be specified to a high enough standard for the systems to be used by operators other than the students who developed the systems. End-user training will have to take place before the inter-operator agreement test can take place.

There is a question as to what data should be included in reliability studies. One point of view is that each performance indicator to be used when presenting results and conclusions must have its level of reliability described. There is another point of view that the observation process should be demonstrated to be reliable and then any performance indicators derived using the method can be deemed reliable through a process of inductive reasoning. This author would not wish to outlaw the second approach as there are sound reasons for using it. The *Journal of Sports Science and Medicine* publishes research in various areas of sports science, including sports performance. The word limit for original research articles is 3,800 including 300 words for each table or figure that is used. This prohibits a complete certification of reliability of every performance indicator to be used in many studies. Students of undergraduate research projects face similar pressures. There may be an overall word limit of around 10,000 words, which can allow students to devote more words to the description of the reliability study in the methods chapter, but this may be at the expense of using words in the discussion for which there are a lot of marks allocated.

The next two sections of this chapter cover the different types of reliability statistic that can be used in performance analysis. These can be used to evaluate intra-operator agreement or inter-operator agreement. The reliability statistic to be used depends on the scale of measurement of the action variable or performance indicator of interest. The four scales of measurement described in statistics textbooks (Vincent, 2005: 6–7) are nominal scale, ordinal scale, interval scale and ratio scale. The first two scales are categorical scales of measurement as variables use values from a finite set of named values. The difference between nominal scale variables and ordinal scale variables is that the values of a nominal scale variable are just different values (for example male and female), whereas ordinal scale variables use values that have a defined order (for example 'never', 'seldom', 'sometimes', 'often' and 'always'). In performance analysis, we further subdivide nominal scales into two different types of nominal scale. The first is where any error involving two values is equally serious. The second type of nominal scale is where there are some pairs of values that are 'neighbouring' and others that are not. An example of a nominal variable with neighbouring values is area of the playing surface. This is not an ordinal variable, but there will be some pairs of areas (or cells) of the playing surface that border each other and others that do not. Where an action occurs on the border of two cells, one operator may record one of the cells and the other operator might record the other cell. This is a less serious error than when an error involves two cells that are not even bordering each other. Computerised systems such as SPSS merge interval and ratio scales into a single numerical scale (called 'scale' in SPSS). This has little impact on the use of descriptive and inferential statistics for the main purpose of a performance analysis project. However, interval and ratio scale variables do need to be considered differently for the purposes of reliability assessment, as some reliability statistics that can be used with ratio scale variables may not be valid for use with interval scale measures. Therefore, there are five scales of measure that we consider in performance analysis research:

1. nominal scale with no neighbouring values;
2. nominal scale with neighbouring values;

3. ordinal scale;
4. interval scale;
5. ratio scale.

RELIABILITY STATISTICS FOR CATEGORICAL VARIABLES

Reliability of nominal scale variables

The reliability of nominal scale variables will be illustrated using an example from a study of elite tennis strategy (Brown and O'Donoghue, 2008b); the variable of interest is point type. The data used in the inter-operator agreement study were from 1,356 points played in 12 different Grand Slam singles matches with at least one men's match and at least one women's match from each of the four Grand Slam tournaments. Hughes *et al.* (2004) proposed the use of a percentage error calculation to evaluate the reliability of nominal variables that computes a single percentage error value. However, in Table 7.1, the current author has also shown the percentage errors for the frequency of each nominal value. The method used by Hughes *et al.* (2004) is shown in equation 7.1 and expresses the sum of the absolute difference in frequencies as a percentage of the sum of the mean frequencies recorded by the two operators (f1 and f2):

$$\text{per cent error} = \Sigma \mid f1 - f2 \mid /(\Sigma \, (f1 + f2)/2) \qquad (7.1)$$

The total percentage error in this example is 4.9 per cent, which is within the typical stated level of error of 5 or 10 per cent. However, as Table 7.2 shows, this total percentage error conceals the fact that there are two values of point type with percentage errors greater than 10 per cent. One of the problems with using percentage error as a reliability statistic is that it may conceal errors that cancel each other out because only the totals are used. For example, Table 7.2 reveals that there were 690 points that both operators agreed were baseline rallies. There were 12 points that the first operator classified as points where the server went to the net and the second operator classified as baseline rallies. When inspecting the total frequency of each point type recorded by the two operators, we see that these 12 errors were cancelled out by 12 of the 33 points that the second operator classified as points where the server went to the net and that the first operator classified as baseline rallies.

The kappa statistic (Cohen, 1960) is a measure of reliability for nominal variables that can be used when we have a chronologically ordered sequence of values for each operator. Where we have accumulated totals through a

Table 7.1 Percentage error of point type in tennis in an inter-operator agreement test (data from Brown and O'Donoghue, 2008b)

Point Type	Operator 1	Operator 2	Absolute Difference	Mean Value	Percentage Error
Ace	79	80	1	79.5	1.3
Double Fault	42	41	1	41.5	2.4
Serve Winner	237	243	6	240	2.5
Return Winner	19	19	0	19	0.0
Server to Net	169	151	18	160	11.3
Receiver to Net	92	78	14	85	16.5
Baseline Rally	718	744	26	731	3.6
Total	1356	1356	66	1356	

Table 7.2 Inter-operator reliability table for point type (data from Brown and O'Donoghue, 2008b)

First Operator	Second Operator							
	Ace	DF	S.W	R.W	S.Net	R.Net	BL	Total
Ace	78						1	79
DF	1	38		1			2	42
S.W	1		232	1	1		2	237
R.W		1		17			1	19
S.Net			2		131	3	33	169
R.Net					7	70	15	92
BL		2	9		12	5	690	718
Total	80	41	243	19	151	78	744	1356

Key:
DF   Double Fault
S.W   Serve Winner
R.W   Return Winner
S.Net   Server attacks net first
R.Net   Receiver attacks net first
BL   Baseline rally

tallying system, the cross-tabulation shown in Table 7.2 cannot be produced. Kappa determines the proportion of points that are agreed by the two operators excluding the proportion that one would expect to be agreed by chance. This is a very important consideration where one value (such as baseline rally) is dominant. If the first operator realised the most common point type was baseline rally and guessed this for every point, the agreement with the second operator would still be greater than 50 per cent of points

(744/1,356). In our example, there were 1,256 of the 1,356 points where the operators agreed on point type (92.6 per cent). Therefore, the proportion of points where the operators agreed on point type, $p_0 = 0.926$. However, this includes many points where the operators could expect to agree by guessing. For example, the fraction of points that the first operator classifies as aces is 79/1,356. Therefore, we could expect this fraction of the 80 points where the second observer classified the point as an ace to be agreed by guessing. Applying this logic to each of the seven point types allows the number of points expected to be agreed by guessing to be determined as shown in Table 7.3.

This gives a proportion that could be expected to be agreed by guessing, $p_C = 466.7/1,356 = 0.344$. If we consider p_0 to be itself divided by one, the proportion of points agreed can be adjusted by subtracting the expected proportion agreed by guessing, p_C, from the top and bottom of this division to give kappa, κ. Cohen's (1960) original kappa statistic is shown in equation 7.2. The equations 7.3 and 7.4 define the proportion of agreements, p_0, and the expected proportion of agreements by chance, p_C, respectively. In these equations, V is the number of different nominal values for the performance indicator of interest and N is the total number of cases recorded by each observer. A is a two-dimensional table of $V \times V$ values representing the number of occasions the observers recorded each value, with $A_{i,j}$ representing the number of occasions that the first observer recorded the ith nominal value and the second observer recorded the jth nominal value.

$$\kappa = \frac{p_0 - p_C}{1 - p_C} \quad (7.2)$$

$$p_0 = \frac{\displaystyle\sum_{k=1}^{V} A_{k,k}}{N} \quad (7.3)$$

$$p_C = \frac{\displaystyle\sum_{k=1}^{V} \left(\frac{\left[\displaystyle\sum_{i=1}^{V} A_{i,k}\right]\left[\displaystyle\sum_{j=1}^{V} A_{k,j}\right]}{N} \right)}{N} \quad (7.4)$$

Table 7.3 Frequency distribution of point type in tennis in an inter-operator agreement test (data from Brown and O'Donoghue, 2008b)

Point Type	Calculation	Expected agreements by guessing
Ace	80 x 79/1356	4.7
Double Fault	41 x 42/1356	1.3
Serve Winner	243 x 237/1356	42.5
Return Winner	19 x 19/1356	0.3
Server to Net	151 x 169/1356	18.8
Receiver to Net	78 x 92/1356	5.3
Baseline Rally	744 x 718/1356	393.9
Total		466.7

Table 7.4 Interpretation of kappa values (Altman, 1991: 404)

Kappa	Strength of agreement
$\kappa \geq 0.8$	Very good
$0.6 \leq \kappa < 0.8$	Good
$0.4 \leq \kappa < 0.6$	Moderate
$0.2 \leq \kappa < 0.4$	Fair
$\kappa < 0.2$	Poor

Therefore, in this example of point type, the kappa value for inter-operator agreement of point type was $(0.926 - 0.344)/(1 - 0.344) = 0.888$. This is a very good strength of agreement according to the Altman's (1991: 404) evaluation scheme for kappa shown in Table 7.4.

Reliability of nominal scale variables with neighbouring values

Very often in performance analysis, the area of the playing surface in which an event occurs is recorded. Hughes and Franks (2004c) described examples of systems where the area of the playing surface where events are performed is recorded for field hockey and squash. Hughes and Franks (2004d) provided further examples of systems where the playing surface is divided into cells for tennis, basketball, soccer and netball. Figure 7.1 illustrates how the netball shooting circle is decomposed for the purposes of analysis (Robinson and O'Donoghue, 2007).

Where a player shoots from an area on the border between areas 2 and 5, one observer might record area 2 while the other might record area 5. Kappa treats this disagreement as a total disagreement the same way as it

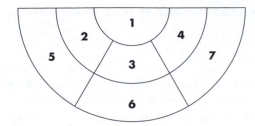

Figure 7.1 Areas of shooting circle of a netball court

would treat a disagreement between areas 2 and 7, which do not border each other. Where the areas recorded by two observers border each other, the observers should receive some credit. This can be done using Cohen's (1968) weighted kappa statistic, which introduced the two-dimensional array of V x V weights. This array is referred to as W. Equation 7.2 is still used to determine κ using p_0 and p_C, but p_0 and p_C are calculated differently using equations 7.5 and 7.6 respectively.

$$p_0 = \frac{\sum\limits_{i=1}^{V} \sum\limits_{j=1}^{V} A_{i,j} W_{i,j}}{N} \tag{7.5}$$

$$p_0 = \frac{\sum\limits_{i=1}^{V} \sum\limits_{j=1}^{V} \left(\frac{\left[\sum\limits_{i=1}^{V} A_{i,k}\right]\left[\sum\limits_{j=1}^{V} A_{k,j}\right] W_{i,j}}{N} \right)}{N} \tag{7.6}$$

In Robinson and O'Donoghue's (2007) example, the weighted version of kappa actually resulted in a slightly reduced value for kappa with the weights that they used. The calculation of the weighted version of kappa is illustrated in the following section, where it also applies.

Reliability of ordinal scale variables

Ordinal variables have all of the properties of nominal variables, but additionally values on an ordinal scale are ordered. An example of an ordinal variable is aggressiveness with which a shot is played in tennis (Boffin, 2004). Table 7.5 shows that Boffin used seven different levels of aggressiveness and used the numbers 1 to 7 to rate them. This variable is clearly

Table 7.5 An ordinal scale for rating aggressiveness of tennis shots (adapted from Boffin, 2004)

Rating	Guidelines
1	Totally defensive. Shot has no depth, spin or power. Barely clears net. Hit out of position on stretch, only aim to get ball in play somehow.
2	Defensive lob. High over net with no pace or spin, but element of control allowing player time to get back into central court position, hit on stretch.
3	Shot hit with limited pace, may have spin, landing about the service line, often hit from behind the baseline, allowing opponent to move forward and be aggressive.
4	Basic neutral rally shot. Landing centrally in court with no attempt to manoeuvre opponent. Lands midway between service line and baseline, mid-pace, will have spin, hit from behind the baseline.
5	Shot played from about the baseline with the aim of moving opponent around. Hit into space away from opponent, either drop shot or ball aimed into open court.
6	Hit with increased pace. Player aiming for shots with lower margin for error, either very deep towards the baseline, very close to the sidelines (literally an inch or two from the lines) or aggressive short angles pushing opponent fully out of court. Usually hit whilst moving forward.
7	Totally aggressive shot. Looking directly for outright winner, low over net, away from opponent or flat out at opponent. Hit as a result of opponents defensive shot (1, 2 or 3) therefore player usually mid-court when hitting shot.

ordinal with increasing levels of aggressiveness from ratings 1 to 7. However, just because the interval between the numbers used is 1, does not mean that this is an interval scale measure. The numbers 1 to 7 merely represent the different values of the scale.

Table 7.6 shows the cross-tabulation of 60 shots rated by two independent operators of Boffin's (2004) system. There are 44 agreements out of 60 shots analysed (p_0 = 0.733) with the proportion where we could expect agreement by guessing to be p_C = 0.164. Therefore kappa value is 0.681, indicating a good strength of inter-operator agreement.

The unweighted version of kappa (Cohen, 1960) is a particularly harsh measure of reliability for this type of variable because minor disagreements (like ratings of 6 and 7 being made by the operators) are treated the same as serious disagreements (like ratings of 1 and 7 being made by the operators). In ordinal scales, every adjacent pair of values is a pairing of neighbouring values. This should be recognised by giving some credit where minor disagreements have occurred between the two operators. Therefore, weights such as those shown in Table 7.7 should be used within Cohen's (1968) weighted kappa statistic for evaluating the reliability of ordinal variables.

Table 7.6 Inter-operator reliability of rating of aggressiveness of tennis shot (fictitious data using Boffin's, (2004) method)

Operator 1 rating	Operator 2 rating							
	1	2	3	4	5	6	7	Total
1	2	1						3
2		3		1				4
3			8	3				11
4				7	2	1		10
5					8	4		12
6					1	7	2	10
7						1	9	10
Total	2	4	8	11	11	13	11	60

This increases p_0 to 51.0/60 = 0.850, p_C to 18.58/60 = 0.310, leading to an increased kappa value of 0.783, which still represents a good strength of agreement between the operators.

Reliability statistics for ratio scale variables

Correlation coefficients are measures of relative reliability that reflect how well values maintain their ranks during reliability studies. This can either be during a test-retest reliability study in physiology, for example, or during independent observations of the same performances in a performance analysis

Table 7.7 Weightings to be used when applying the weighted kappa statistic to inter-operator reliability of a shot aggressiveness in tennis

Operator 1	Operator 2						
	1	2	3	4	5	6	7
1	1.0	0.5					
2	0.5	1.0	0.5				
3		0.5	1.0	0.5			
4			0.5	1.0	0.5		
5				0.5	1.0	0.5	
6					0.5	1.0	0.5
7						0.5	1.0

reliability study. Relative reliability studies should not be used on their own because they do not describe error values. Measures of absolute reliability are concerned with the amount of error expressed in the units of measurement or as a percentage of the value recorded. Measures of absolute reliability that are used with numerical scale variables are mean absolute error, root mean square error, percentage error, standard error of measurement, 95 per cent limits of agreement and 95 per cent ratio limits of agreement.

There are two numerical scales of measurement of interest to performance analysis: the interval scale and the ratio scale. Interval scales of measurement have all of the properties of ordinal scales except that in addition to the values being ordered, there is a fixed interval between values. Therefore, subtraction is meaningful when using interval scales whereas only comparison (greater than and less than comparison) can be used with ordinal scales. Displacements, velocities, accelerations, angular velocities and angular accelerations are all examples of interval scale variables that can be used in performance analysis. Some of these can have negative values as well as positive values, meaning that division is not always meaningful. Ratio scale variables possess all of the properties of interval scale variables except that they also use zero to represent a total absence of value. Therefore, division and ratios of values are meaningful. Distance, time, height and mass are measured on ratio scales, with 6m (or s or kg) being twice as long (or heavy) as 3m (or s or kg). Ratios and division such as this cannot be done validly with variables measured on interval scales that are not also ratio scales. Ratio scale variables are more common in performance analysis research then interval scale variables and, therefore, the author has chosen to describe reliability statistics that can be used with ratio scale variables before describing reliability statistics for interval scale variables.

In this section, 100m split times during an 800m track athletics event are used as an example of reliability evaluation for ratio scale variables. The data are the split times recorded during Brown's (2005) study of the 2004 Olympic women's 800m. The final of the women's 800m was analysed by two independent operators using Brown's manual system. Once elapsed times are recorded, split times can be determined for each athlete during each 100m section of the race. Figures 7.2 and 7.3 plot the data recorded by the two operators against each other for elapsed times and split times respectively. There are a total of 64 value pairs as there were eight split times for each of the eight athletes participating in the race. If Pearson's r is used as a measure of relative reliability (how well the values maintain their rankings within the set of 64 values), the use of elapsed times gives an artificially high value of r > 0.999. Basically, the elapsed times include eight 100m timings, eight 200m timings, and so on, up to eight 800m timings. Essentially, there are eight different variables, whose values form clusters within the scatter plot shown in Figure 7.2.

Figure 7.3 plots the split times recorded by the two operators against each other. There are a total of 64 value pairs as there were eight split times

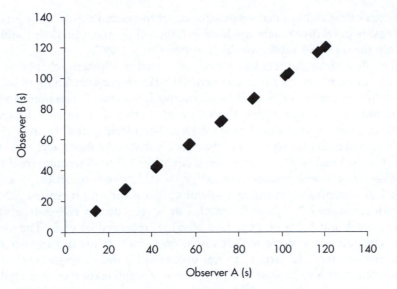

Figure 7.2 Relative reliability of elapsed time in 800m running (using data from Brown 2005)

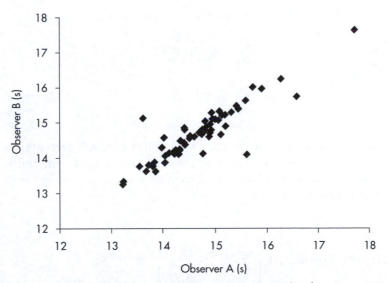

Figure 7.3 Relative reliability of split time in 800m running (using data from Brown 2005)

for each of the eight athletes participating in the race. Pearson's r is used as a measure of relative reliability (how well the values maintain their rankings within the set of 64 values) of the split times (r = 0.897).

The data from the reliability study are used to illustrate change in the mean, standard error of measurement (SEM), 95 per cent limits of agreement, percentage error as well as mean absolute error. Percentage error is determined by expressing the absolute error between recorded values as a percentage of the mean of the two values. The mean is used because there may be no reason to expect one observer's value to be more accurate than the other's. Hughes *et al.* (2004) used equation 7.7 to determine total percentage error where multiple frequency variables were recorded. In equation 7.7, |x| represents the magnitude of x; that is the sign is removed. Note that in equation 7.7, A_i and B_i are *i*th values recorded by two independent operators *A* and *B* out of a total on N values recorded by each. The use of this equation can be criticised because it does not actually use any percentage errors in the calculation of total percentage error. A single total of all absolute errors is expressed as a percentage of a single total of mean values between the two observers. Therefore, a high frequency variable such as passes will dominate the divisor of this equation and conceal large percentage errors for other frequency variables such as tackles, crosses and shots. The percentage error should be determined for these frequency variables individually.

$$Total\ \%\ Error = 100\ \frac{\sum_{i=1}^{N} |A_i - B_i|}{\sum_{i=1}^{N} (A_i + B_i)/2} \qquad (7.7)$$

Percentage error values can be combined where they are percentage error values from the same variable. For example, 100m split time during an 800m race could be combined to determine a mean percentage error, as shown in equation 7.8.

$$Mean\ \%\ Error_1 = 100 \left[\sum_{i=1}^{N} \frac{|A_i + B_i|}{(A_i + B_i)/2} \right] /N \qquad (7.8)$$

One criticism of mean percentage error 7.8 is that it divides the absolute difference by the mean of the two values and can thus determine low values for percentage error. The problem with this is that the full range of values from 0s to the mean is not the meaningful range of values that would be observed for 100m split times; an athlete will not run any 100m section of an 800m race faster than 10s and will certainly not cover 100m in 0s. An alternative is to specify a theoretical maximum value, *Max*, of say 18s, as well as a theoretical minimum, *Min*, and express the inter-operator disagreement as a percentage of the range of meaningful values from *Min* to *Max*, as in equation 7.9.

$$Mean\ \%\ Error_3 = 100 \left[\sum_{i=1}^{N} \frac{|A_i + B_i|}{Max - Min} \right] / N \quad (7.9)$$

Equations 7.8 and 7.9 are used in Table 7.8 to determine percentage error yielding 64 percentage values in each case. The first column of Table 7.9 identifies the athlete and point in the race at which the time was recorded. The next two columns are the elapsed times up to each point in the race by each athlete according to observers *A* and *B* respectively. The split times (columns 4 and 5) can be determined by subtracting the elapsed time for the previous 100m point in the race from any given 100m point from 200m to 800m. This can easily be accomplished in Microsoft Excel. Split times are used in the reliability study because, a) using elapsed times would conceal errors, b) split times are all 100m times whereas elapsed times use totally different ranges of values at the different 100m points in the race, and c) the analytical goals of the study usually involve split times rather than elapsed times.

The final two columns of Table 7.8 report the two different types of percentage error described in the current chapter. Column 9 is the unadjusted percentage error value used in equation 7.8, which expresses the absolute error as a percentage of the mean of the values recorded by the two observers (column 8 shows the mean values). The final column of the table shows the percentage error produced by equation 7.9, which does not use the mean of the two operators' values. The error is expressed as a percentage of the meaningful range of values; in this case the meaningful range of values is 8s, the difference between a theoretical minimum of 10s and a theoretical maximum of 18s.

We can see that the mean of the percentage errors is 1.26 per cent and 2.32 per cent, but we may wish to summarise the percentage errors in different ways. For example, we may wish to report the maximum percentage errors, which are 10.44 per cent and 19.00 per cent. Alternatively, we may wish to use the 95th percentile of the percentage errors to give a percentage error that we can state that 95 per cent of percentage errors are less than. This can be done with the Percentile function in Microsoft Excel and in the current example the results would be 4.34 per cent and 7.83 per cent.

In computing mean absolute error, the difference between the two operators' recorded split times is first computed (column 6 of Table 7.8) and then

Table 7.8 Calculation of mean percentage error for split time (data from Brown, 2005)

Athlete: Distance	Elapsed Time A (s)	Elapsed Time B (s)	Split Time A (s)	Split Time B (s)	Difference (s)	Absolute Difference (s)	Mean Value (s)	%Error (equ 7.8)	%Error (equ 7.9)
1:100m	13.81	13.76	13.81	13.76	-0.05	0.05	13.785	0.36	0.63
1:200m	27.54	27.56	13.73	13.80	0.07	0.07	13.765	0.51	0.88
1:300m	41.75	41.74	14.21	14.18	-0.03	0.03	14.195	0.21	0.38
1:400m	56.43	56.45	14.68	14.71	0.03	0.03	14.695	0.20	0.38
1:500m	71.84	71.91	15.41	15.46	0.05	0.05	15.435	0.32	0.63
1:600m	86.97	87.10	15.13	15.19	0.06	0.06	15.160	0.40	0.75
1:700m	103.24	103.33	16.27	16.23	-0.04	0.04	16.250	0.25	0.50
1:800m	120.95	120.95	17.71	17.62	-0.09	0.09	17.665	0.51	1.13
				There are a total of 64 rows (8 athletes x 8 splits)					
2:100m	14.42	14.37	14.42	14.37	-0.05	0.05	14.395	0.35	0.63
2:200m	28.46	28.42	14.04	14.05	0.01	0.01	14.045	0.07	0.13
8:600m	87.31	87.89	15.59	15.62	0.03	0.03	15.605	0.19	0.37
8:700m	103.89	103.62	16.58	15.73	-0.85	0.85	16.155	5.26	10.63
8:800m	119.62	119.62	15.73	16.00	0.27	0.27	15.865	1.70	3.37
Mean			14.69	14.69	0.00	0.19	14.690	1.26	2.32
SD			0.78	0.74	0.35	0.29	0.74	1.98	3.64

the magnitude of this is determined with the ABS function in Microsoft Excel (column 7). The mean of the 64 absolute errors is the mean absolute error, which is 0.19s. The use of the magnitude of the error guards against errors cancelling each other out in the calculation of the reliability statistic.

The absolute errors may be reported in the units of measurement (s in this case) without using percentages. This may be appropriate when relating measurement error to the analytical goals of the study. Mean absolute error, the 95th percentile for absolute errors or maximum absolute error could be reported. There may be particular splits that are more reliable than others due to camera positioning or track markings. For example, Brown and O'Donoghue (2007) found that the split times at 300m, 700m, 1,100m and 1,500m points of the 1,500m where the athletes passed the finish line on each lap were more accurate than the 100m, 500m, 900m and 1,300m split times where the athletes were passing the 200m start. Therefore, reliability statistics reported separately for different points in the race could be a more informative way of presenting reliability results. An example of this for the 800m is shown in Table 7.9.

The first seven columns of Table 7.8 can also be used in the calculation of 95 per cent limits of agreement, change of the mean and standard error of measurement. One of the disadvantages of using mean absolute error or a derivative of it is that an overall error statistic is reported without providing any information about the different components of error. Atkinson and Nevill (1998) described how total error is the sum of systematic bias and random error. Systematic bias is concerned with any tendency for one operator to record a higher value than another. For example, in an inter-operator agreement study, one operator may classify a movement as being performed at a high intensity more often than another operator. If we are aware of such a tendency, then this can be addressed when data are gathered by the operators. This could be done by using the difference between the observer means and adding half of this value to any value recorded by the observer who tends to record lower values and subtract half of this value from any value recorded by the observer who tends to record higher values.

Random error is the additional error that may be due to individual perceptual errors, data entry errors or operator fatigue. Bland and Altman's (1986) 95 per cent limits of agreement are computed using the individual errors (column 6 of Table 7.8). There is very little systematic bias here with the mean difference in the split times recorded by the two operators being 0.00s. As we can see in column 6 of Table 7.8, not all errors are the same as the systematic bias value. This is because random error also has an influence on measurement. Random error is computed using the standard deviation of the differences between the two operators' recorded split times. This can be done using the STDEV function in Microsoft Excel, giving 0.346s. The standard deviation only represents 68 per cent of the spread of error values if the errors are normally distributed. Therefore, the standard deviation has to be multiplied by $z_{1-0.05/2}$ = 1.96 to give a spread of 95 per cent of the error values. In this example, the random error is ±0.678s. The 95 per cent limits of agreement are expressed as the systematic bias ± the random error (0.000 ± 0.678s). The Bland Altman plot (Figure 7.4) can be a useful way of illustrating the types of errors associated with different values. However, decisions about the level of reliability can be made from the systematic bias ± the random error. One thing that researchers must remember is that the random error is not a mean error or a typical error value, but the error value that only 5 per cent of errors are expected to exceed.

Hopkins (2000) recommended using change of the mean and standard error of measurement (SEM) to represent systematic bias and random error respectively. The change of the mean is simply the difference between the mean values recorded by the two operators (14.69s - 14.69s = 0.0s). This will always give the same result as the systematic bias in Bland and Altman's (1986) 95 per cent limits of agreement. The measure of random error recommended by Hopkins (2000) is the SEM (sometimes referred to as the 'typical error'). This is computed by dividing the standard deviation of the error values by √2, which covers 52 per cent of the errors if they are normally distributed. In Brown's (2005) data of 100m split times during the 800m, the SEM is 0.247s.

Table 7.9 summarises the absolute reliability of the split times recorded using different reliability statistics. The most reliable split time was 500m to 600m according to mean absolute error and the two percentage error statistics. These three measures do not distinguish between the systematic bias and random error components of total error. The change in the mean and the systematic bias are equivalent. Knowledge of systematic bias allows measurements to be adjusted for known tendencies of particular observers,

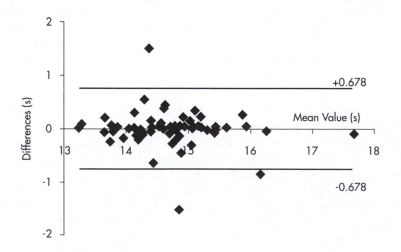

Figure 7.4 Bland Altman Plot (using data from Brown 2005)

Table 7.9 Summary of absolute reliability statistics (s)

Race section	Mean Absolute error	%Error (equ 7.8)	%Error (equ 7.9)	Change in the mean	SEM	Systematic Bias	Random Error
0m-100m	0.13	0.93	1.64	0.04	0.16	0.04	0.44
100m-200m	0.11	0.78	1.33	0.00	0.10	0.00	0.28
200m-300m	0.29	2.04	3.66	0.22	0.38	0.22	1.06
300m-400m	0.30	2.06	3.80	-0.27	0.38	-0.27	1.04
400m-500m	0.15	0.99	1.88	0.13	0.09	0.13	0.26
500m-600m	0.10	0.67	1.25	0.06	0.12	0.06	0.33
600m-700m	0.28	1.85	3.52	-0.27	0.23	-0.27	0.63
700m-800m	0.12	0.76	1.47	0.09	0.10	0.09	0.27
All	0.19	1.26	2.32	0.00	0.24	0.00	0.68

meaning that only the random error remains. SEM and random error are both derived from the standard deviation of the inter-operator errors.

The use of the 95 per cent limits of agreement is based on the assumption of homoscedasticity. This means that there should be no association between the magnitude of the errors (column 7 of Table 7.8) and the mean values (column 8). In this case the data is homoscedastic (r = 0.040). Heteroscedasticity is where there is a positive association between the magnitude of the errors and the mean values recorded by the two operators. That is, higher values are associated with higher errors than lower values are; this is sometimes termed a 'shotgun' effect. Where such a 'shotgun' effect occurs, the systematic bias and random error are better expressed using ratios (x/÷) than using differences (+/-) because larger values have larger errors. Readers who are interested in the detail of how 95 per cent ratio limits of agreement are calculated are directed to the worked example by Atkinson and Nevill (1998).

Reliability statistics for interval scale variables

There are different types of interval scale variables in performance analysis that include negative values. Examples from analysis of technique include displacement, velocity, acceleration, angular velocity and angular acceleration. While automatic data capture systems such as Coda (CodaMotion, Rothley, Leicestershire, UK) and Vicon (Vicon, Los Angeles, CA) are used in many kinematic analysis studies, there are other systems with analysis tools that involve human operator input. These systems include Dartfish (Dartfish, Fribourg, Switzerland), siliconcoach (Siliconcoach, Dunedin, New Zealand) and Quintic biomechanics (Quintic Consultancy Ltd, Coventry, UK). Where human operator input is involved in the calculation of angles, distances or any velocity and acceleration derivatives of these, it is essential to demonstrate the reliability of the methods used.

Many of the reliability statistics that can be used with ratio scale variables can also be used with interval scale variables. Change of the mean, standard error of measurement, mean absolute error, root mean square error and 95 per cent limits of agreement can all be used with both interval and ratio scale variables. These techniques are based on differences in values recorded by different operators. Subtraction of values is valid on any interval or ratio scale.

Percentage error, as defined by Hughes *et al.* (2004), is one technique that cannot be used with interval scale variables that include negative values. During the calculation of percentage error, the absolute difference between two operators' recorded values is divided by the mean of the two values. This could involve a division by zero error or a division by a negative number leading to a negative percentage error being computed. Even when total percentage error is used, the sum of the absolute errors is divided by a sum of mean values which may be positive, negative or zero. The adjusted percentage error (equation 7.8) can be used to evaluate the reliability of interval scale variables because the absolute errors are always divided by a positive value representing the range of meaningful values. The 95 per cent ratio limits of agreement should not be used with interval scale variables. This reliability statistic involves dividing one operator's recorded value by the corresponding value recorded by the other operator. Such divisions can only be validly applied to ratio scale data.

SUMMARY

Notational analysis has often been portrayed as a purely objective analysis technique. However, this chapter has explained that when a human operator is part of the system used to measure performance variables, there will always be an element of subjective judgement when classifying behaviour. It is, therefore, essential to provide guidance on system use, train system users and determine the level of reliability of the system. Different variables are measured on different measurement scales requiring the use of different reliability statistics.

CHAPTER 8

ANALYSIS OF QUANTITATIVE SPORTS PERFORMANCE DATA

INTRODUCTION

This chapter describes quantitative data analysis, how it should be linked to the research question and the interpretation of results. The selection of data analysis techniques to help answer different types of research question is covered. Performance analysis data often go through various stages of processing before statistical tests can be applied. There are spreadsheet processing techniques, such as pivot tables, that can be used to determine performance indicator values from the raw data. There are many different ways of designing performance analysis investigations with independent samples, related samples, associations between performance indicators or combinations of these. This chapter describes the analysis of such data using parametric and nonparametric inferential tests and how they are undertaken using SPSS (SPSS Inc., Chicago, Il). The assumptions of the various tests and how these can be tested are also covered. The chapter includes examples of performance data that are analysed in SPSS and the SPSS output information to be used in the production of the results.

DESIGN ISSUES

There are many types of performance analysis investigation and indeed investigations that involve performance analysis in conjunction with other disciplines. This means that there are no standard experimental or observational designs that can be used, and some investigations may be the first to use a particular research design. Therefore, an ability to understand when different statistical tests should be applied, how they should be used, how tests can be tailored for particular research designs and how statistical results can be presented are essential for the successful performance analysis researcher. The analysis depends on the number of independent factors involved in the study, the types of samples of data involved, the scale of measurement and distributions of the dependent variables and above all, the purpose of the study.

The purpose of the study should be understood before collecting data in quantitative research. Specifying formal hypotheses to be tested allows the gathering and analysis of data to be driven by the goal of the study which is to answer the research question. As explained in Chapter 4, hypotheses are specified in terms of the variables that are used in the study. It is, therefore, necessary to have an understanding of the variables involved in the study as well as their scales of measurement. Where categorical variables are used to form samples, the types of sample also need to be understood when forming hypotheses and when conducting the eventual analysis. This chapter will now briefly recap on the different types of sample that can be compared. Samples can be independent samples or related samples. Independent samples are samples of data drawn from different participants or matches that are being compared; examples are data from male or female participants or from players of different positional roles. Because the samples contain independent data, it is not necessary for the size of the samples to be the same. Related samples are samples of data that are related to the same participants or matches; examples are player performances when playing at home and when playing away from home, or player performances in the four different quarters of a netball match. It is necessary for related samples to contain the same number of values because they are matched samples that represent different conditions for the same subjects.

The purpose of the study could be to test the relationship between variables or it could be to test for differences between samples. Figure 8.1 shows the different types of test that can be used for these different purposes when different types of samples are used. The decision tree shown in Figure 8.1 is not exhaustive and only represents those tests of associations that are between two variables and tests of differences involving single independent factors. The tests of association may involve variables with different scales of measurement. For example, we may wish to test the relationship between a nominal variable and an ordinal variable. In such a situation we should use the chi-square test of independence because an ordinal variable is also a nominal variable (with ordered values) but a nominal variable is not always an ordinal variable. Similarly, if we wished to test the association between an ordinal variable and a numerical variable (interval or ratio scale measure), we would use either Kendall's τ or Spearman's ρ because numerical variables are also ordinal variables but not all ordinal variables are numerical variables.

Figure 8.1 presents two different tests of difference between samples in each of the four situations shown. The decision as to whether to use the parametric test or the alternative nonparametric test depends on the scale of measurement of the dependent variable and its distribution. These tests of differences can only be used with ordinal and numerical scale variables to test differences between the medians and the means respectively. Where the dependent variable is an ordinal scale measure, a nonparametric test should be used. Nonparametric tests use the rankings of values rather than the values themselves when calculating the test result. Ordinal values can be transformed into ranks in the same way as numerical values can be transformed into ranks.

Where the dependent variable is measured on a numerical scale, the decision to use a parametric test or a nonparametric test depends on whether the data satisfy the assumptions of the test. Parametric tests are called parametric tests because the results are calculated using the sample parameters (means, standard deviations and sample sizes). They are more powerful

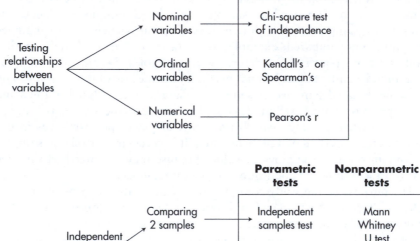

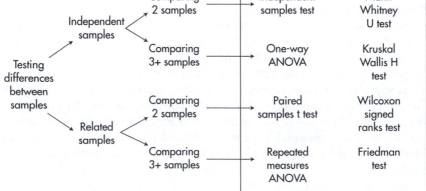

Figure 8.1 Decision tree of statistical tests

than nonparametric alternatives that lose information when transforming values into ranks. Therefore, where researchers can validly use parametric or nonparametric tests, they should always choose to use the parametric test. There are three main assumptions of parametric tests, although some tests have additional assumptions:

1. the dependent variable is measured on an interval or ratio scale;
2. the dependent variable is normally distributed; and
3. the variance of the dependent variable is similar between the different samples being compared.

The first of these assumptions has already been addressed as we are only considering the use of parametric tests for numerical scale dependent variables. Simple inspection of the definition of the dependent variable will determine whether it is a numerical scale variable or not. There are tests for the normality of the dependent variable and tests for equality of variances, which will be outlined later in this chapter. Where a numerical variable passes these tests, parametric tests can be validly applied to the data. Where the dependent variable is not normally distributed or where the samples being compared have differing variances, it is necessary to use an alternative nonparametric test. The nonparametric tests are 'distribution free', meaning they can be used with any data that can be ranked. There are some assumptions that apply to nonparametric tests; for example the Kruskal Wallis H test requires that there are at least five values in each sample being compared.

The same research question can be examined using different research designs. This often depends on the unit of analysis being used in the study. For example, we may wish to compare home and away performances (using some dependent performance indicator) in a sport by using any of the following designs:

*   Comparing home and away performances within a given set of matches. Here, the unit of analysis is the match and we would be comparing home and away team performances as related conditions within matches. A paired samples t-test could be used if the data satisfied the necessary assumptions. The performance of a team is influenced by the performance of the opponent, so in this research design it would not be appropriate to treat home and away performances as independent samples.
*   Comparing teams' mean performances in matches where they play at home and where they play away from home. Here, the unit of analysis is the team and the home and away conditions are used to form related samples with each team's mean home match performance and mean away match performance being compared. Again a paired samples t-test could be used if the data satisfied the necessary assumptions.

- Comparing a set of home performances with a set of away perform-ances where different sets of teams are involved in the home perform-ances and the away performances. Here the unit of analysis is the match performance, as the same team may be involved in multiple matches within the sample of home performances as well as multiple matches within the sample of away performances. The two samples would be considered as two independent sets of performances being compared. Therefore, an independent samples t-test could be used if the data satis-fied the assumptions of the test.

- Comparing the proportion of events that are performed positively at home with the proportion that is performed positively away from home. In this example, there may not be enough matches to use the perform-ance indicator of interest. For example, our performance indicator might be the percentage of passes that are successful and we may only have data from four matches involving a total of eight teams. A way of analysing the data is to make the event the unit of analysis rather than whole performances or matches. The events are classified using a nominal variable of two values (positive and negative) and the propor-tion of positively performed events could be compared between home and away performances using a chi-square test of independence.

These different ways of investigating the same research question mean that the researcher should consider the volume of data to be gathered in each case, the feasibility of being able to access this data, and how much time and effort would be required. Once a particular research design is chosen, the researcher develops the system to gather the data required and may have to process the data in order to produce the specific values to be entered into a statistical analysis package.

DESCRIPTIVE AND INFERENTIAL STATISTICS

Descriptive statistics are used to summarise samples in terms of given vari-ables. Nominal variables can be summarised using frequency distributions showing how many subjects (participants, events or matches) belong to each defined category. The mode is a useful descriptive statistic that can be used with nominal or ordinal variables. Ordinal variables can be summarised using the median as well as the lower and upper quartiles. Where the ordinal scale variable has very few values, frequency distributions can also be used. Where numerical variables are used, samples can be described using the mean (or median) as a measure of location. However, we will also be inter-ested in the spread of values around the mean (or median). Therefore, the standard deviation (or inter-quartile range) should also be used to describe the sample. Descriptive statistics can be represented in the text of the results section or as tables or charts. Typically, a combination of all three methods of presentation is used. Chapter 10 covers the write up of different types of research report.

Inferential statistics are used to determine whether the differences between samples are statistically significant. Rather than relying on a subjective judgement of the difference between two sample means (or medians or modes), an inferential test uses precise criteria to determine whether the difference between two samples is statistically significant. Similarly, inferential statistics can be used to determine if the relationship or association between two variables is statistically significant. Inferential tests produce a probability, p, which is used to determine if there is a significant difference (or relationship) or not. Most research studies are based on samples rather than whole populations of interest. Therefore, any difference between samples may not be representative of a difference between the corresponding populations. This is called sampling error, and it affects all research based on samples. There are two types of error that can occur that are shown in Table 8.1.

The probability, p, is the probability of making a Type I error (claiming that the difference between samples represents a difference between corresponding populations when in reality it does not). An inferential test does not look at the whole population and so there is always a chance that the conclusions of the test do not reflect what would have been concluded from a census of the whole population. This is why sentences in research papers show the p value at the end, as in the following example:

There was a significantly greater percentage of net points played in men's singles tennis than in women's singles tennis (p < 0.001).

The p value being less than 0.001 is the acknowledgement that the result is not a definite proof of a difference in general and there is a small probability (< 0.001) that a Type I error has been made. We rarely know if a Type I error has actually been made because results for the wider population are unknown and so all we can do is present the probability that a Type I error has occurred.

The lower the p value is, the higher the significance. This is because p is the probability of making a Type I error. The lower the probability of

Table 8.1 Type I and Type II Errors

Conclusion of Test (Samples)	Reality (Populations) unknown	
	Difference	No Difference
Difference	OK	Type I error
No Difference	Type II error	OK

making such a mistake, the greater the significance. Usually in social science research, p values of less than 0.05 are deemed to be significant. There are two ways of reporting p values: they can either be reported as they are to a chosen number of decimal places or they can be classified as $p < 0.05$, $p < 0.01$ or $p < 0.001$ (significant) or $p > 0.05$ or $p = 0.05$ (indicating no significant difference or relationship). The 0.05 threshold value of p for concluding a significant result is not universal and there are different threshold values that can be chosen. The choice of p value that indicates significance depends on the upper limit that the researcher wants to place on the chance of making a Type I error. This in turn will be influenced by the potential risks of a mistake and the consequences involved. For some studies one chance in 20 of making a Type I error may be intolerably high while in other studies researchers may choose to use a threshold p value of 0.1.

Students find inferential statistics harder to do than descriptive statistics but this does not make inferential statistics more important than descriptive statistics. Reporting inferential statistics without descriptive statistics is worse than reporting descriptive statistics without inferential statistics. Descriptive statistics tell us how long, how far, how many, what percentage and so on. These values can be compared with known norms, standards and results reported in previous research. The p values produced by inferential tests are important but not as important as the descriptive statistics; the inferential statistics are the 'icing on the cake' but we need the cake first and foremost. If we were to conclude that the percentage of net points is significantly different between men's singles tennis matches and women's singles tennis matches, we do not know, a) which type of match has the higher percentage of net points, and b) the actual percentage of points in each case. It could be that the percentage of net points is high (or low) in both types of game with one game having a significantly greater percentage than the other. Alternatively, it could be that one type of game has a high percentage of net points and the other type of game has a low percentage of net points with the difference being significant. Descriptive statistics are needed to provide this information and then inferential statistics can determine if any differences are significant or not.

TESTS OF RELATIONSHIPS

Relationships between nominal variables

In this chapter some of the same examples will be used as they were when describing the use of hypotheses in Chapter 4. The association between two nominal variables is tested using a chi-square test of independence and the example used to illustrate this is the association between court surface and match outcome. This test operates by comparing the observed frequencies with the expected frequencies if the distribution of cases between the columns of the cross-tabulation as a whole applied to each row equally.

Recall from Chapter 4 that there are two classes of match outcome: matches where the higher ranked player won and matches that were upsets. The example in Chapter 4 is extended to use three court surfaces: grass, clay and cement. There are a total of 404 matches, each being represented by a row within an SPSS data sheet. In SPSS we go to the 'Analyse' menu, and then the 'Descriptive Statistics' submenu and select 'Crosstabs'. The convention used in this book for this navigation through the menus in SPSS is *Analyse → Descriptive Statistics → Crosstabs*. A popup window appears that allows us to set up a cross-tabulation of frequencies with court surface as the row of the table and match outcome as the column. We can use the 'cells' button to request percentages per row as well. This information is of assistance when inspecting the table if there are differing numbers of matches played on each surface. Clicking on the 'Continue' button completes the selecting of percentages. To include the result of a chi-square test of independence with the cross-tabulation of frequencies, we click on the 'Statistics' button and place a tick in the 'Chi-square' field. Clicking on 'Continue' to complete the use of the 'Statistics' popup window and 'OK' to complete the 'Crosstabs' analysis commences the calculation of the statistical results by SPSS. The results appear in the output viewer, which is a separate window to the data sheet used to store the data being analysed. The output consists of two tables, as shown in Figure 8.2.

The percentage of upsets occurring in clay court matches is greater than the percentage occurring on the other two surfaces. However, there is no significant relationship between court surface and match outcome as the p value of the chi-square test (Asymp. Sig. in the first row of the second table of Figure 8.2) is greater than 0.05. The student should extract the results required from these tables and produce the following summary inferential and descriptive statistics:

> There were 42 upsets out of 129 matches on clay (32.6%) which was a greater proportion than the 37 out of 134 on grass (27.6%) and the 30 out of 141 on cement (21.3%). However, there was no significant relationship between court surface and match outcome ($\chi^2_2 = 4.4$, p = 0.111).

The chi-square result presented within parentheses could have read (p = 0.111), (p > 0.05), ($\chi^2_2 = 4.4$, p > 0.05) or ($\chi^2_2 = 4.4$, p = 0.111). The author prefers to use the fourth way as it shows the chi-square value and the p value. When the p value is classified as greater than 0.05 information is lost and the reader will not be aware if the test was close to being significant or not. In reporting the chi-square result, the subscripted value 2 is the number of degrees of freedom for the test and 4.4 is the chi-square value presented to 1 decimal place. The degrees of freedom are computed by multiplying (rows − 1) x (columns − 1). We have a 3 x 2 cross-tabulation here and so there are 2 degrees of freedom.

Surface * Outcome Crosstabulation

			Outcome		Total
			To Form	Upset	
Surface	Grass	Count	97	37	134
		% within Surface	72.4%	27.6%	100.0%
	Clay	Count	87	42	129
		% within Surface	67.4%	32.6%	100.0
	Cement	Count	111	30	141
		% within Surface	78.7%	21.3%	100.0%
Total		Count	295	109	404
		% within Surface	73.0%	27.0%	100.0%

Chi-Square Tests

	Value	df	Asymp. Sig. (2-sided)
Pearson Chi-Square	4.393[a]	2	.111
Likelihood Ratio	4.433	2	.109
Linear-by-Linear Association	1.455	1	.228
N of Valid Cases	404		

[a] 0 cells (.0%) have expected count less than 5.
The minimum expected count is 34.80.

Figure 8.2 Cross-tabulation of frequencies and chi-square test of independence

There is an assumption for the chi-square test that at least 80 per cent of cells in the cross-tabulation must have an expected frequency of at least five. Where this is not the case, some rows or columns can be merged for the purpose of the chi-square test but still shown separately in the descriptive results (Thomas *et al.*, 2005: 184).

If the cross tabulation is a 2 x 2 table with a single degree of freedom, we still report the chi-square value. However, the p value we report must be the two-tailed p value from the 'Fisher's Exact Test' row of results in SPSS rather than 'Pearson's Chi-squared' row of results. The results of Fisher's exact test are only shown when the cross-tabulation is a 2 x 2 table.

Relationships between numerical variables

Pearson's coefficient of correlation, r, is used to determine if there is an association between numerical variables. A correlation coefficient will be between -1.0 and +1.0. The sign of the correlation represents the direction of the relationship: negative means that as one variable gets larger the other gets smaller whereas a positive correlation means that as one variable gets larger so does the other. The magnitude of the correlation represents the strength of association between the two variables: a magnitude close to 0.0 means that there is little or no relationship between them whereas a correlation with a magnitude close to 1.0 is a very strong relationship. To determine the coefficient of correlation, r, in SPSS use *Analyse → Correlate → Bivariate* and enter the variables of interest into the 'Variables' area. The example used here is the relationship between the percentage of first serves that are played in during 252 Grand Slam tennis matches, the percentage of points won when the first serve is in and the percentage of points won when a second serve is required. Although we have entered three variables, SPSS still performs bivariate correlations using each pair of the variables that were entered. The output from SPSS is shown in Figure 8.3.

The correlations of 1 are where variables are correlated against themselves and the results below the top left to bottom right diagonal of cells are a mirror image of those above. Therefore, there are only three correlations; one for each pair of variables. There is a negative relationship between the percentage of first serves played in and the percentage of points won when the first serve was in ($r = -0.375$, $p < 0.001$). Note that a p value of 0.000 to three decimal places is shown as $p < 0.001$ rather than 0.000. The more the first serve is in, the less likely it is that the point will be won when it is in. There is no relationship between first serves played in and percentage of points won on second serve ($r = +0.010$, $p = 0.881$) and a weak positive relationship between the percentage of points won on first serve and second serve ($r = +0.174$, $p = 0.006$). The author prefers not to use p values with correlations as the coefficient of determination, r^2, is a better measure of the meaningfulness of the correlation because it represents the proportion of one variable that is explained by other. For example, 0.141 (or 14.1 per cent) of the value for the percentage of points won when the first serve is in

Correlations

		%1st Serves In	% Won on 1st Serves	% Won on 2nd Serves
% 1st Serves In	Pearson Correlation	1	-.375**	.010
	Sig. (2-tailed)		.000	.881
	N		252	252	252
% Won on 1st Serves	Pearson Correlation	-.375**	1	.174**
	Sig. (2-tailed)	.000		.006
	N	252	252	252
% Won on 2nd Serves	Pearson Correlation	.010	.174**	1
	Sig. (2-tailed)	.881	.006	
	N	252	252	252

** Correlation is significant at the 0.01 level (2-tailed).

Figure 8.3 Pearson's coefficient of correlation

is explained by the percentage of points when the first is in. The remaining 85.9 per cent of the value for the percentage of points won when the first serve is in is explained by other factors such as the quality of ground stroke technique and ability at the net.

SPSS allows scattergrams to be drawn to illustrate the relationship between two variables. We could determine the equation of the line of best fit relating a pair of variables using *Analyse → Regression → Linear* with some variable (Y) as the dependent variable and some variable (X) as the independent variable. The table of regression coefficients provides B_0 and B_1, which are the intercept of the line and the Y axis and the gradient of the line respectively. These coefficients are used in the equation of the line of best fit, which is in the form $Y = b_0 + b_1.X$. Note that it is possible to produce a regression equation for Y in terms of more than one X variable. In the case of the tennis data, the correlations are not large enough to justify the use of linear regression.

Relationships between ordinal variables

There are two correlation techniques that can be used with any variables that can be transformed into ranks; these are Spearman's ρ (rho) and Kendal's τ (tau), which also produce values between -1.0 and +1.0. Spearman's ρ is used if there are at least 20 pairs of values being analysed, otherwise Kendall's τ is recommended. To determine the values of these correlations use *Analyse → Correlate → Bivariate* and enter the variables of interest into the 'Variables' area. There is an area of the popup entitled 'Correlation Coefficients'; the user should make sure Pearson's r is unticked and Spearman's ρ and Kendall's τ are ticked. The same example is used here and the SPSS output is as shown in Figure 8.4.

THE ASSUMPTIONS OF PARAMETRIC TESTS

There are three main assumptions of parametric statistical tests of difference between samples:

1. The dependent variable is measured on an interval or ratio scale.
2. The dependent variable is normally distributed.
3. The variance of the dependent is similar between the different samples being compared.

The scale of measurement of a variable can be determined by inspecting the definition of the variable. Therefore, the assumption of normality and the assumption of homogeneity of variances are covered using the mean duration of a rally in a tennis match as an example. There are 252 matches from Grand

Correlations

			%1st Serves In	% Won on 1st Serves	% Won on 2nd Serves
Kendall's tau_b	% 1st Serves In	Correlation Coefficient	1.000	-.254**	-.050
		Sig. (2-tailed)	.	.000	.243
		N	252	252	252
	% Won on 1st Serves	Correlation Coefficient	-.254**	1.000	.171**
		Sig. (2-tailed)	.000	.	.000
		N	252	252	252
	% Won on 2nd Serves	Correlation Coefficient	-.050	.171**	1.000
		Sig. (2-tailed)	.243	.000	.
		N	252	252	252
Spearman's rho	% 1st Serves In	Correlation Coefficient	1.000	-.371**	-.069
		Sig. (2-tailed)	.	.000	.274
		N	252	252	252
	% Won on 1st Serves	Correlation Coefficient	-.371**	1.000	.245**
		Sig. (2-tailed)	.000	.	.000
		N	252	252	252
	% Won on 2nd Serves	Correlation Coefficient	-.069	.245**	1.000
		Sig. (2-tailed)	.274	.000	.
		N	252	252	252

** Correlation is significant at the 0.01 level (2-tailed).

Figure 8.4 Nonparametric correlations

Slam tennis tournaments in the dataset, with 116 of them being women's singles matches and 136 of them being men's singles matches.

Tests of normality

There are different tests of normality; in this book, the use of the Kolmogorov-Smirnov test and the Shapiro-Wilks test are preferred. The Shapiro-Wilks test is used if there are fewer than 50 values in the overall sample. If there are 50 or more values then the Kolmogorov-Smirnov test is used. In our example, there are 252 tennis matches and so the Kolmogorov-Smirnov test is used. This test (like the Shapiro-Wilks test) compares the distribution of values with a perfect normal distribution of the same mean and standard deviation as the set of values. A p value of under 0.05 means that the distribution is significantly different to the required normal distribution and that the data fail to satisfy the assumption of normality. In SPSS, the Kolmogorov-Smirnov test is done by using *Analyse → Descriptive Statistics → Explore* and entering the variable rally length into the 'Dependent List'. The 'Plots' button activates a popup where we can ask for 'Normality plots and tests'. This will provide the results of both the Kolmogorov-Smirnov test and the Shapiro-Wilks test. Figure 8.5 shows the output of the tests of normality, which is just one of several pieces of output to be produced. However, all we need to know is that the p (Sig. value) of the Kolmogorov-Smirnov test is less than 0.05 and so the data have violated the assumption of normality.

Levene's test of homogeneity of variances

Levene's test of homogeneity of variances compares the variances of the samples involved in the study and produces a p value of under 0.05 if these variances are significantly different. A p value of 0.05 or greater means that the assumption of equal variances is satisfied by the data. In SPSS, we cannot obtain the results of a homogeneity test without actually undertaking a parametric test. We can either use an independent samples t-test (if there are two samples) or a one-way ANOVA test (if there are three or more samples). Where we have related samples, it is necessary to temporarily reorganise the data so they appear to be independent samples just to get the results of Levene's test.

In SPSS, the independent samples t-test is done and the output is ignored if the data fail Levene's test. The independent samples t-test is done in SPSS with *Analyse → Compare Means → Independent Samples t-test* with the dependent variable (rally length) being placed in the 'Test variables' area and the independent variable (gender) being placed in the 'Grouping Variable' area. The grouping variable should use numeric codes to represent the values (for example 1 for female and 2 for male). These can be labelled with the appropriate text strings and if the user has entered the data as text rather than numeric codes, this can be overcome through *Transform → Automatic Recode*. When defining the groups within the independent samples t-test, the appropriate codes are used (for example 1 and 2) and then the test can be executed using the 'OK' button. There are two tables in the output, shown in Figure 8.6, but we are only concerned with the second table. This contains two sets of columns representing Levene's test of homogeneity of variances and the independent samples t-test. The p value of Levene's test of homogeneity of variances is 0.065, which is greater than 0.05 meaning that there is no significant difference between the sample variances and that the assumption of approximately equal variances is satisfied. Where the data fail Levene's test, the independent samples t-test can still be used with a reduced number of degrees of freedom. The first row of t-test results are used if the data pass Levene's test with the

Tests of Normality

	Kolmogorov-Smirnov[a]			Shapiro-Wilk		
	Statistic	df	Sig.	Statistic	df	Sig.
Rally Length	.090	252	.000	.937	252	.000

[a] Lilliefors Significance Correction

Figure 8.5 Tests of normality

Group Statistics

	gender	N	Mean	Std. Deviation	Std. Error Mean
Rally Length	female	116	7.223	2.3217	.2156
	male	136	5.242	1.7464	.1498

Independent Samples Test

		Levene's Test for Equality of Variances		t-test for Equality of Means							
									95% Confidence Interval of the Difference	
		F	Sig.	t	df	Sig. (2 tailed)	Mean Difference	Std. Error Difference	Lower	Upper
Rally Length	Equal variances assumed	3.436	.065	7.717	250	.000	1.9812	.2567	1.4756	2.4869
	Equal variances not assumed			7.548	210.932	.000	1.9812	.2625	1.4638	2.4986

Figure 8.6 Independent samples t-test

second row of results potentially being used where Levene's test is failed. In performance analysis research, the convention is that both the test of normality and the test of homogeneity of variances should be passed to justify the use of parametric procedures.

If we have three or more samples we would obtain the results of Levene's test using a one-way ANOVA through *Analyse → Compare Means → One-way ANOVA*. The independent variable (surface for example) is placed into the 'Factor' area and the dependent variable (rally length) is placed in the 'Dependent List'. Unlike the independent samples t-test, SPSS does not provide the results of Levene's test by default for the one-way ANOVA. Therefore, the 'Options' button is used to access the 'Options' popup window where we can ask for 'Homogeneity of variance test'.

PARAMETRIC TESTS

Independent samples t-test

The independent samples t-test is used to compare samples drawn from different independent units of analysis (matches, participants or performances). The example of comparing mean rally duration between men's and women's singles tennis matches is used to illustrate the use of the independent samples t-test. The independent samples t-test is done in SPSS with *Analyse →Compare Means → Independent Samples t-test* with the dependent variable (rally length) being placed in the 'Test variables' area and the independent variable (gender) being placed in the 'Grouping Variable' area and defined. The output provided by SPSS for the independent samples t-test contains two tables, as shown in Figure 8.6. The first table shows the means and the standard deviations of the two samples. The second table contains two sets of columns representing Levene's test of homogeneity of variances and the independent samples t-test. The results of the independent samples t-test are augmented to the descriptive statistics for the two samples and can be presented in text form as follows:

> The mean rally duration of 7.2±2.3s in women's singles matches was significantly longer than the 5.2±1.7s in men's singles matches (t_{250} = 7.7, p < 0.001).

The subscripted 250 is the degrees of freedom for the independent samples t-test (the combined sample size minus two) and 7.7 is the t value expressed to one decimal place. The means and standard deviations could be represented as bars and error bars respectively if these results were to be presented graphically, or the results could be presented as a table. Of course, due to the failure of the data to satisfy the assumption of normality, we would be unable to use the independent samples t-test in this example and would instead use the non-parametric alternative, which is the Mann Whitney U test.

Paired samples t-test

The paired samples t-test is used to compare two different conditions related to the same group. We will use the example of 252 tennis matches where we wish to compare the percentage of points won by the serving player under two conditions: when the first serve is in and when a second serve is required. In SPSS, the paired samples t-test is done using *Analyse → Compare Means → Paired samples t test* selecting the pair of variables of interest. These have to be selected as a pair, which appears in the 'Current Selection' area, before the variables are transferred as a pair into the 'Paired Variables' area. Clicking on the 'OK' button provides the results shown in Figure 8.7.

The results are reported as follows:

> Players won 67.4±9.2% of points on first serve, which was significantly greater than the 48.3±10.5% of points won on second serve (t_{251} = 23.8, p < 0.001).

The means and standard deviations could be presented within a table or within a bar graph with bars representing the two means and error bars representing the two standard deviations.

Paired Samples Statistics

		Mean	N	Std. Deviation	Std. Error Mean
Pair 1	%Won on 1st serves	67.382	252	9.1782	.5782
	%Won on 2nd serves	48.286	252	10.5399	.6640

Paired Samples Correlations

		N	Correlation	Sig.
Pair 1	%Won on 1st serves & %Won on 2nd serves	252	.174	.006

Paired Samples Test

	Paired Differences							df	Sig. (2-tailed)
	Mean	Std. Deviation	Std. Error Mean	95% Confidence Interval of the Difference		t			
				Lower	Upper				
Pair 1 %Won on 1st serves − %Won on 2nd serves	19.0959	12.7147	.8010	17.5185	20.6734	23.841	251		.000

Figure 8.7 Paired samples t-test

One-way ANOVA

ANOVA stands for 'analysis of variances tests', which describes what these tests do. An ANOVA test will determine the variance between different samples and the variance within samples. For there to be a significant difference between samples, one would expect the variance between samples to be noticeably greater than the variance seen within samples. ANOVA tests use the F-ratio, which is the ratio of between-samples variance to within-samples variance when analysing the data. An F-ratio of one or less can never be significant because it indicates that there is no more variance between samples than there is within samples.

If there is a significant difference between the samples, it would be desirable to investigate differences between individual pairs of samples. This can be done using post hoc tests. The author prefers the Bonferroni adjusted post hoc test because it intentionally adjusts the p value to avoid inflating the probability of making a Type I error. For example, if the threshold p value used to identify significance is 0.05 and if there are four samples being compared, there will be six pairs of samples meaning that the chance of a Type I error occurring within the analysis could be $1 - 0.95^6 = 0.26$ rather than 0.05. The Bonferroni adjustment ensures that the chance of making a Type I error during the post hoc tests does not inflate above 0.05 or whatever threshold p value is being used.

To do a one-way ANOVA in SPSS we use *Analyse → Compare Means → One-Way ANOVA,* which is used here to compare the mean rally length at the four different Grand Slam tennis tournaments. The variable 'surface' is our independent factor that we have hypothesised as influencing rally length. The descriptive statistics do not appear by default and have to be selected through the 'Options' button, while the 'Post-Hoc' button is used to select one of a variety of alternative post hoc procedures; in this case we will use the Bonferroni adjusted post hoc test. The SPSS output is shown in Figure 8.8.

The t-tests only require a single value for degrees of freedom because we know we are comparing two samples. However, with an ANOVA test we need to represent the number of samples as well as the total number of subjects. The first degree of freedom (Between Groups) is one less than the number of samples. The second degree of freedom (Within Groups) is the number of values minus the number of samples. The F-ratio is 36.3 when expressed to one decimal place. The means and standard deviations are taken from the first table of Figure 8.8, with the details of the ANOVA test and post hoc tests coming from the second and third tables respectively. It should be noted that there are six post hoc tests rather than 12, as the post hoc test results show each post hoc test twice, once in each order. The results could be reported in text as follows:

> We find rally lengths of 6.3±1.6s at the Australian Open, 8.2±2.4s at the French Open, 4.6±1.6s at Wimbledon and 5.8±1.9s at the US Open. In this case, surface had a significant influence on rally length ($F_{3,248}$ = 36.3, p < 0.001) with Bonferroni adjusted post hoc tests revealing that rallies were significantly longer at the French Open than at all other tournaments (p < 0.001), significantly shorter at Wimbledon than at the Australian (p < 0.001), French (p < 0.001) and US (p < 0.01) Opens with no significant difference between rally lengths at the Australian and US Opens (p > 0.05).

Repeated measures ANOVA

The repeated measures ANOVA test is used to compare three or more related samples. The repeated measures ANOVA is done in SPSS through *Analyse → General Linear Model → Repeated Measures.* An example of a dependent variable is the percentage of time spent performing high intensity activity in a game of netball and an example of an independent variable is the match quarter. There are 28 players and we wish to determine if there is a difference in the percentage of time spent performing high intensity activity between the four quarters. However, in SPSS we don't have a variable quarter or a single dependent variable for the percentage of high intensity activity performed in a quarter. We have four repeated measurements of our

Descriptives

Rally Length

	N	Mean	Std. Deviation	Std. Error	95% Confidence Interval for Mean		Minimum	Maximum
					Lower Bound	Upper Bound		
Synthetic	62	6.288	1.5852	.2013	5.885	6.690	3.6	12.0
Clay	57	8.156	2.4088	.3191	7.517	8.795	4.2	15.7
Grass	66	4.632	1.5999	.1969	4.238	5.025	2.5	9.9
Cement	67	5.828	1.9193	.2345	5.359	6.296	3.0	13.0
Total	252	6.154	2.2559	.1421	5.874	6.434	2.5	15.7

ANOVA

Rally Length

	Sum of Squares	df	Mean Square	F	Sig.
Between Groups	389.623	3	129.874	36.282	.000
Within Groups	887.733	248	3.580		
Total	1277.356	251			

Multiple Comparisons

Dependent Variable: Rally Length
Bonferroni

(I) surface	(J) surface	Mean Difference (I–J)	Std. Error	Sig	95% Confidence Interval	
					Lower Bound	Upper Bound
Synthetic	Clay	−1.8683*	.3472	.000	−2.792	−.945
	Grass	1.6559*	.3346	.000	.766	2.546
	Cement	.4601*	.3334	1.000	−.427	1.347
Clay	Synthetic	1.8683*	.3472	.000	.945	2.792
	Grass	3.5242*	.3421	.000	2.614	4.434
	Cement	2.3284*	.3409	.000	1.422	3.235
Grass	Synthetic	−1.6559*	.3346	.000	−2.546	−.766
	Clay	−3.5242*	.3421	.000	−4.434	−2.614
	Cement	−1.1958*	.3281	.002	−2.068	−.323
Cement	Synthetic	−.4601	.3334	1.000	−1.347	.427
	Clay	−2.3284*	.3409	.000	−3.235	−1.422
	Grass	1.1958*	.3281	.002	.323	2.068

*. The mean differrence is significant at the .05 level.

Figure 8.8 One-way ANOVA

conceptual dependent variable in the datasheet (HI_Q1 to HI_Q4). Therefore we must follow these steps to do a repeated measures ANOVA in SPSS:

1. We state that we have a within-subjects factor called 'Quarter' which is measured at four levels. This within-subjects' factor is entered into our ANOVA model using the 'Add' button.

2.   We identify the four repeated measurements for the percentage of time spent performing high intensity activity as HI_Q1 to HI_Q4; our 'with-in-subjects variables'.

3.   SPSS does not actually allow post hoc tests to be set up for within-subjects' factors using the 'Post Hoc' facility. Instead we have to use 'Options' where we can transfer 'Quarter' to the 'compare means for' area. We place a tick in 'compare main effects' and use the Bonferroni adjusted post hoc test.

4.   Before leaving the 'Options' popup window, we ask for descriptive statistics to be provided because these are not provided in the standard output for the test.

The repeated measures ANOVA has an additional assumption, sphericity, which is homogeneity of variance and covariance between the repeated measures. In this case, because Mauchly's test of sphericity gives a p value of greater than 0.05, we can continue to use the test looking at the 'Sphericity Assumed' results within the 'Tests of Within Subjects Effects' table within the output shown in Figure 8.9. Where the assumption of sphericity is violated, there is an increased chance of making a Type I error if we have a significant result. Therefore, the process of selecting which row of results to use from the SPSS output is as follows:

•   If sphericity is satisfied (Mauchly's test having a p values of greater than 0.05), then the Sphericity Assumed results are used.

•   If sphericity is violated, but the result of the Sphericity Assumed ANOVA test is not significant (p > 0.05), then this should be used. This is because it is impossible to make a Type I error when we are concluding no significant difference and using a strict condition would actually increase our chance of making a Type II error.

•   If sphericity is violated, but the results of the Greenhouse Geisser adjusted ANOVA are significant (p < 0.05) then the Greenhouse Geisser results should be reported. The Greenhouse Geisser is a strict condition where we can be more confident in any significant result found.

•   Where we have violated sphericity and the strictest condition (Greenhouse Geisser) reveals no significant difference, but the most liberal condition (Sphericity Assumed) does reveal a significant difference, then the Huynh Feldt ANOVA results should be used.

Figure 8.9 does not show all of the output provided by SPSS, just the tables from which we can extract those figures needed to produce the results.

The results can be presented in text as follows:

The percentage of time spent performing high intensity activity in the four quarters was $23.0 \pm 5.3\%$, $22.0 \pm 5.4\%$, $20.7 \pm 5.8\%$ and $19.2 \pm 5.4\%$. There was a significant difference between the four quarters ($F_{3,81}$ =

7.2, p < 0.001) with Bonferroni adjusted post hoc tests revealing that players spent a significantly lower percentage of time performing high intensity activity in the last quarter than the first (p < 0.05) and second (p < 0.01).

Factorial ANOVA

The four tests previously described are used to test the difference between different levels of a single factor. For example, gender is a factor measured

Descriptive Statistics

	Mean	Std. Deviation	N
hi_q1	23.0475	5.28167	28
hi_q2	22.0082	5.37944	28
hi_q3	20.6757	5.77453	28
hi_q4	19.2214	5.40273	28

Mauchley's Test of Sphericity[b]

Measure: MEASURE _1

Within Subjects Effect	Mauchley's W	Approx. Chi-Square	df	Sig.	Epsilon[a] Greenhouse -Gessier	Huynh -Feldt	Lower -Bound
quarter	.714	8.682	5	.123	.816	.904	.333

Tests the null hypothesis that the error covariance matrix of the orthonormalized transformed dependent variables is proportional to an identity matrix.
a. May be used to adjust the degrees of freedom for the averaged tests of significance.
   Corrected tests are displayed in the Tests of Within-Subjects Effects table.
b. Design: Intercept
   Within Subjects Design: quarter

Tests of Within-Subjects Effects

Measure: MEASURE _1

Source		Type III Sum of Squares	df	Mean Square	F	Sig.
quarter	Sphericity Assumed	231.007	3	77.002	7.239	.000
	Greenhouse-Geisser	231.007	2.449	94.344	7.239	.001
	Huynh-Feldt	231.007	2.711	85.210	7.239	.000
	Lower-Bound	231.007	1.000	231.007	7.239	.012
Error (quarter)	Sphericity Assumed	861.659	81	10.638		
	Greenhouse-Geisser	861.659	66.111	13.034		
	Huynh-Feldt	861.659	73.198	11.772		
	Lower-Bound	861.659	27.000	31.913		

Figure 8.9 Repeated measures ANOVA

at two levels, surface of a tennis court was a factor measured at four levels, service in tennis was measured at two levels (first serve and second serve) and quarter of a netball match was measured at four levels. There are occasions where we wish to analyse two factors at the same time, especially if the combination or interaction of these factors has an influence on the given dependent variable above the effect of the factors in isolation. If we were comparing the activity of boys and girls but also comparing children of two different age groups, it is possible that we might not find a significant gender or age effect. This is because the variability within each age group caused by having children of two different genders in each group may result in the independent samples t-test result not being significant. Similarly, the variability within each gender group caused by having children of two different age groups in each gender group may also result in the independent samples t-test result not being significant. The independent t-test requires a large enough difference between the two sample means in relation to the pooled standard deviation of the samples and the number of values included in the test. A two-way ANOVA, on the other hand, could include both gender and age group, testing each factor while controlling for known differences between the values of the other factor. This would be more likely to produce significant results in the example described. A further advantage of a two-way ANOVA is that it is possible that a significant interaction might be found between the two factors.

An example of a significant interaction is in a study of 72 soccer players who were classified according to level (elite, semi-professional or amateur) and playing position (centre back, wing back, midfield or forward). A time-motion study measured the percentage of time spent performing high inten-

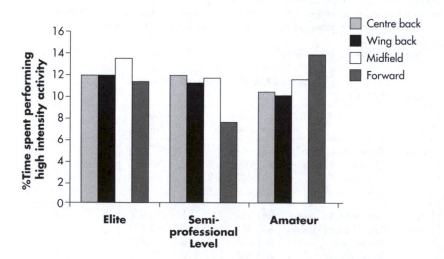

Figure 8.10 A significant interaction between two factors (fictitious example of percentage match time spent performing high intensity activity by soccer players)

sity activity for these players. Figure 8.10 shows the descriptive results that are produced.

To perform a two-way ANOVA with two between-subjects factors in SPSS, we use *Analyse → General Linear Model → Univariate*. We transfer the percentage time spent performing high intensity activity into the 'Dependent Variable' area and include both level and playing position as fixed factors. The 'Options' popup is used to request descriptive statistics and 'Post Hocs' popup is used to choose post hoc tests. Figure 8.11 shows that p values are determined for the effect of level, position, and the interaction of these two factors (level * position). Neither level nor playing position has a significant effect. However, the interaction of these two factors is significant ($F_{6,60}$ = 2.4, p = 0.037) because, as Figure 8.11 shows, the percentage of time spent performing high intensity activity depends on both level and playing position. The players who perform the greatest amount of high intensity activity are the elite midfielders and the amateur forwards, while the players who perform the least high intensity activity are the semi-professional forwards and the amateur centre backs and wing backs.

Two way ANOVA tests can also be performed on two within-subjects factors (the within-within design) or on one within-subjects factor and one between-subjects factor (a mixed design). Both of these tests would be set up in the repeated measures ANOVA facility of SPSS. It is also possible to have factorial ANOVA tests that involve more than two factors, but the more factors that are included the more data are required for the study. This is a quantitative analysis chapter within a book on research methods for performance analysis and it is not possible to describe all of the possible variations of ANOVA test that can be done. However, readers requiring more detail on factorial ANOVA tests are directed to Chapter 11 of Vincent's (2005) textbook and Chapter 3 of Ntoumanis's (2001) textbook.

Tests of Between-Subjects Effects

Dependent Variable: pc_hi

Source	Type III Sum of Squares	df	Mean Square	F	Sig.
Corrected Model	229.503ᵃ	11	20.864	1.882	.060
Intercept	8784.154	1	8784.154	792.288	.000
level	39.920	2	19.960	1.800	.174
pos	20.637	3	6.879	.620	.605
level * pos	160.447	6	26.741	2.412	.037
Error	665.225	60	11.087		
Total	10392.100	72			
Corrected Total	894.728	71			

a. R Squared = .257 (Adjusted R Squared = .120)

Figure 8.11 Two-way factorial ANOVA results

NONPARAMETRIC TESTS

Mann Whitney U test

The Mann Whitney U test is the nonparametric alternative to the independent samples t-test. It is used to compare two independent samples where the assumptions of parametric procedures have been violated. The Mann-Whitney U is done in SPSS with *Analyse → Nonparametric → 2 Independent Samples,* entering the grouping variable and the test variable as before. We will use the same example as used for the independent samples t-test: the effect of gender on the mean rally duration in Grand Slam tennis matches. Figure 8.12 shows that the output contains a table of mean ranks and a table summarising the Mann Whitney U test results.

When we use this to compare the mean rally duration between male and female players we find a significant difference ($U = 3981.0$, $z = -6.8$, p < 0.001). As we are comparing two groups of matches, the test transforms the 252 values into ranks; 1 for the match with the lowest mean rally duration and 252 for the match with the highest mean rally duration. If there was no significant difference, the mean rank for each group would be about 126.5. Here, the mean rank for female matches is 160.12 compared with 97.8 for male matches. This indicates that rally durations are significantly shorter in men's singles than women's singles. To complete the results we need the medians of the two samples for mean rally duration; this is obtained using *Analyse → Compare Means → Means,* transferring gender into the 'Independent List' and rally length into the 'Dependent List'. The 'Options' button is used to access a popup window where we can transfer 'median' into the cell statistics area. The output from SPSS is shown in Figure 8.13. This allows us to express the results as follows:

Ranks

	gender	N	Mean Rank	Sum of Ranks
Rally Length	female	116	160.18	18581.00
	male	136	97.77	13297.00
	Total	252		

Test Statistics[a]

	Rally Length
Mann-Whitney U	3981.000
Wilcoxon W	13297.000
Z	−6.774
Asymp. Sig. (2.tailed)	.000

a. Group Variable: Gender

Figure 8.12 Mann Whitney U test

Report

Median

gender	Rally Length
female	6.620
male	5.049
Total	5.934

Figure 8.13 Descriptive statistics comparing medians between samples

The mean rally duration in women's singles matches (median = 6.6s) was significantly longer than that of men's matches (median = 5.0s) (U = 3981.0, z = -6.8, p < 0.001).

Wilcoxon Signed Ranks test

The Wilcoxon Signed Ranks test is the nonparametric alternative to the paired samples t-test and is used to compare two related samples. The test is accessed in SPSS with *Analyse → Nonparametric → 2 Related Samples*. We use the same example of 252 Grand Slam tennis matches that was used to illustrate the paired samples t-test: the percentage of points won by the server when the first serve is in and when a second serve is required. The SPSS output for the test is shown in Figure 8.14.

The second of the two tables reveals that there is a significant difference between the two conditions, with the first table showing that there were 15

Ranks

		N	Mean Rank	Sum of Ranks
%Won on 2nd serves – %Won on 1st serves	Negative Ranks	237[a]	130.48	30924.00
	Positive Ranks	15[b]	63.60	954.00
	Ties	0[c]		
	Total	252		

a. %Won on 2nd serves < %Won on 1st serves
b. %Won on 2nd serves > %Won on 1st serves
c. %Won on 2nd serves = %Won on 1st serves

Test Statistics[b]

	%Won on 2nd Serves – % Won on 1st Serves
Z	−12.938[a]
Asymp. Sig. (2-tailed)	.000

a. Based on positive ranks
b. Wilcoxon Signed Ranks Test

Figure 8.14 Wilcoxon Signed Ranks test

matches where players won a greater percentage of points on second serve but in the other 237 matches players won a greater percentage of points on first serve. These results should be used in conjunction with the median values that are obtained using *Analyse → Descriptive Statistics → Explore* and entering the two SPSS variables of interest into the 'Dependent List'. The output of descriptive statistics includes the two required medians of 67.7 and 48.9 per cent, as shown in Figure 8.15.

The results can be presented as follows:

The median percentage of points won when the first serve was in of 67.7% was significantly greater than the 48.9% when a second serve was required ($z = -12.9$, $p < 0.001$).

These results could be presented graphically, in a table or in text as above.

Descriptives

			Statistic	Std. Error
%Won on 1st Serves	Mean		67.382	.5782
	95% Confidence Interval for Mean	Lower Bound	66.243	
		Upper Bound	68.521	
	5% Trimmed Mean		67.526	
	Median		67.679	
	Variance		84.239	
	Std. Deviation		9.1782	
	Minimum		37.8	
	Maximum		88.5	
	Range		50.7	
	Interquartile Range		12.1	
	Skewness		−.271	.153
	Kurtosis		−.036	.306
%Won on 2nd Serves	Mean		48.286	.6640
	95% Confidence Interval for Mean	Lower Bound	46.978	
		Upper Bound	49.594	
	5% Trimmed Mean		48.188	
	Median		48.904	
	Variance		111.089	
	Std. Deviation		10.5399	
	Minimum		7.7	
	Maximum		100.0	
	Range		92.3	
	Interquartile Range		10.9	
	Skewness		.595	.153
	Kurtosis		4.666	.306

Figure 8.15 Descriptive statistics comparing two samples

Kruskal Wallis H test

The Kruskal Wallis H test is the nonparametric alternative to the one-way ANOVA test. It is used to compare three or more independent samples and is accessed in SPSS with *Analyse → Nonparametric → K Independent Samples*. This is used here for the same example of Grand Slam tennis as was used to illustrate the one-way ANOVA test. The mean rally duration is compared between the four different tournaments (with surface being distinguished by the labelled codes 1 to 4). Rally length is placed in the 'Test Variable List' and surface is placed in the 'Grouping Variable'. Figure 8.16 shows the resulting output.

There are 252 matches and so if there was no significant difference between the surfaces we would expect a mean rank of about 126.5 for each surface. The Kruskal Wallis H test produces an *H* value that approximates to the chi-square distribution when all samples contain at least five values. Therefore, the chi-square value shown in the SPSS output is actually the *H* statistic of the test. However, this is compared with the chi-square distribution when determining the p value and uses a degrees of freedom value that is one less than the number of groups. Medians are determined in the same way as they are determined for a Mann Whitney U test.

Unfortunately, SPSS does not provide post hoc tests with the Kruskal Wallis H test, but there is a mechanism recommended by Vincent (2005: 250–53) that can be done in Microsoft Excel, or a series of Mann-Whitney U tests could be employed. If a series of Mann Whitney U tests are being done, then an adjusted p value should be applied to avoid inflation of the probability of making a Type I error. If we have K groups, then we have

Ranks

	surface	N	Mean Rank
Rally Length	Synthetic	62	135.31
	Clay	57	188.65
	Grass	66	72.92
	Cement	67	118.25
	Total	252	

Test Statistics[a,b]

	Rally Length
Chi-Square	78.858
df	3
Asymp. Sig.	.000

[a] Kruskal Wallis Test
[b] Grouping Variable: surface

Figure 8.16 Kruskal Wallis H test

N = K x (K - 1)/2 pairs of groups. If we are using a threshold value for p of 0.05, then equation 8.1 shows that the p value to be used in the post hoc Mann Whitney U tests must give an overall probability of avoiding a Type I error in any of the post hoc tests of 0.95.

$$(1-p)^N = 0.95 \qquad (8.1)$$

Therefore, transforming both sides of equation 8.1 into natural logarithms and making p the subject of the equation gives equation 8.2, which evaluates to 0.0085 when N is 6.

$$p = e^{1-\left[\frac{\ln 0.95}{N}\right]} \qquad (8.2)$$

However, usually the Bonferroni adjustment is determined by simply dividing the threshold p value by N. This still gives 0.05/6 = 0.0083 in this case. Using the Kruskal Wallis H tests and the Bonferroni adjusted Mann Whitney U tests along with the medians allows us to describe the results as follows:

> The median for mean rally duration was 6.2s on synthetic surfaces, 7.7s on clay, 4.3s on grass and 5.7s on cement. Surface had a significant influence on the mean duration of rallies within matches (H_3 = 79.8, p < 0.05) with mean rally duration being significantly longer on clay than on all other surfaces (p < 0.05), mean rally duration being significantly shorter on grass than on all other surfaces (p < 0.05) and there being no significant difference between synthetic and cement surfaces for mean rally duration (p > 0.05).

Friedman test

The Friedman test is the nonparametric equivalent of the repeated measures ANOVA test. The test will be illustrated using the same example of the percentage of time spent performing high intensity activity between the four quarters of a netball match that was used to illustrate the repeated measures ANOVA test. We perform the test in SPSS through *Analyse → Nonparametric → K Related Samples*. To compare the percentage of time spent on high intensity activity between the four quarters we simply transfer all four repeated measures of the dependent variable (HI_Q1 to HI_Q4) into the 'Test Variable' area. We click on 'Statistics' and ask for descriptive statistics with quartiles, which will provide the median for each repeated measure (50th percentile). Bonferroni-adjusted Wilcoxon Signed Ranks tests can be used to compare individual pairs of quarters using a similar process to the Bonferonni adjustment of Mann Whitney U tests for the Kruskal Wallis H test. Figure 8.17 shows the SPSS output for the test and the Bonferonni adjusted Wilcoxon signed ranks tests. The results can be reported as follows:

The median for the percentage of time spent performing high intensity activity in the four quarters was 23.4%, 22.6%, 20.0% and 18.6% respectively. The Friedman test revealed significant differences between the four quarters ($\chi^2_3 = 13.3$, $P < 0.01$) with the percentage time spent performing high intensity activity being significantly lower in the fourth than it was in the first quarter ($p < 0.05$) and the second quarter ($p < 0.05$). There was no significant difference between any other pair of quarters ($p > 0.05$).

MULTIVARIATE ANALYSIS TECHNIQUES

Many performance analysis investigations use nonparametric techniques due to data frequently failing to satisfy the assumptions of parametric procedures. However, there are many occasions where parametric procedures are used in performance analysis research and there is a possibility that multivariate analysis techniques could be used as well. The remainder of this chapter will briefly identify different multivariate procedures. Readers who

Descriptive Statistics

	N	Mean	Std Deviation	Minimum	Maximum	Percentiles 25th	50th (Median)	75th
hi_q1	28	23.0475	5.28167	15.07	33.17	18.5450	23.4350	27.4050
hi_q2	28	22.0082	5.37944	12.01	32.33	18.5625	22.6200	25.5150
hi_q3	28	20.6757	5.77453	10.57	33.28	16.2825	19.9950	25.7925
hi_q4	28	19.2214	5.40273	11.31	32.08	15.3300	18.5500	22.6100

Test Statistics[a]

N	28
Chi-Square	13.286
df	3
Asymp. Sig.	.004

a. Friedman Test

Test Statistics[b]

	hi_q2 – hi_q1	hi_q3 – hi_q1	hi_q4 – hi_q1	hi_q3 – hi_q2	hi_q4 – hi_q2	hi_q4 – hi_q3
Z	−1.093[a]	−2.186[a]	−2.756[a]	−1.822[a]	−3.189[a]	−1.936[a]
Asymp. Sig. (2-tailed)	.274	.029	.006	.0.68	.001	.053

a. Based on positive ranks.
b. Wilcoxon Signed Ranks Test.

Figure 8.17 Friedman test and supporting Wilcoxon Signed Rank tests for comparing pairs of related samples

wish to find out more about multivariate analysis techniques are directed to Tabachnick and Fidell's (1996) comprehensive textbook on multivariate statistics as well as Chapter 12 of Vincent's (2005) textbook.

Multivariate ANOVA tests

We have already covered ANOVA tests where there are multiple independent factors. These factorial ANOVA tests are still univariate ANOVA tests because there is only one dependent variable. The multivariate ANOVA (MANOVA) test includes two or more dependent variables. There are many different types of MANOVA depending on the nature and number of independent variables; we can have repeated measures MANOVA tests as well as factorial MANOVA tests. The F-tests for the independent variables are based on an optimal linear composite of the dependent variables. There are alternative versions of the MANOVA test providing different test statistics such as Hotelling's T^2 and Wilk's λ. The MANOVA test is a single test that determines whether or not the independent factor(s) influence the set of dependent variables. It can be seen as a Type I error containment mechanism to be used prior to applying univariate ANOVA tests to individual dependent variables.

Analysis of covariance

A covariate can be used with a univariate ANOVA test as well as multivariate ANOVA tests. A covariate is a variable that has an influence on the dependent variable that we wish to address when analysing the effect of some independent factor. For example, we may be comparing the 60m sprint time of three groups of players who have done different training programmes. Reaction time may have an influence on 60m performance that would distort the results of the study if the players in the different groups had differing reaction times. Reaction time would, therefore, be included as a covariate when comparing the effect of the three training programmes on 60m sprint performance. This test is called an ANCOVA test (analysis of covariance) and can be thought of as a cross between regression analysis and an ANOVA test. The test determines which players have low 60m sprint times for players of their reaction times and compares these differences between observed and expected 60m times in order to adjust for the impact of reaction time. Covariates can also be used within multivariate ANOVA tests; these are called MANCOVA tests where there are at least one independent factor, two or more dependent variables and at least one covariate.

Multiple regression analysis

Linear regression produces an equation for the straight line that best fits a set of points on a two-dimensional graph. Linear regression produces a model for some dependent variable Y in terms of some independent variable X in the form $Y = b_0 + b_1.X$ where b_0 is a constant that represents the intercept with the Y axis and b_1 is the gradient of the line. In multiple linear regression, we have more than one independent variable and the regression line exists in a multidimensional space. Where there are N independent variables, the model for Y takes the form $Y = b_0 + b_1X_1 + ... + b_NX_N$. With any form of regression analysis, users should avoid extrapolating the model beyond the range of any independent variable and only make forecasts for Y based on X values that are within the ranges used to produce the model.

The author uses the example of predicting matches in the 2002 FIFA World Cup (O'Donoghue et al., 2004d) as an example of the dangers of extrapolating beyond the values observed in previous data that were used to create the model. The model represented the fact that teams who travelled further to international soccer tournaments did not perform as well in matches as teams who travelled shorter distances. Because the 2002 FIFA World Cup involved teams travelling further than they had done in previous World Cups, the predictive model based on multiple linear regression forecast that Japan would defeat South Korea in the final. There are other forms of regression analysis besides linear regression, and where the relationship between points is curvilinear, for example, linear regression is not the best method to use. One method used in sports science research is cosinor analysis, which fits the cosine curve of best fit to a set of points. This is particularly useful in the investigation of diurnal variation in sport and exercise (Reilly and Waterhouse, 2009).

Statistical classification

One of the issues of debate in performance analysis research is whether paired samples t-tests should be used to compare the performances of winning and losing teams within matches. The objection to the use of the paired samples t-test is that match outcome is the independent variable when the test is used and we are hypothesising whether the outcome influences the process of performance. It is a case of 'putting the cart before the horse'; the process should influence the match outcome rather than the other way round. Binary logistic regression is a statistical analysis technique that models dichotomous group membership on a set of numerical variables. The two groups could be matches where the higher ranked team won and matches that were upsets (O'Donoghue, 2005b, 2006c, O'Donoghue and Williams, 2004, O'Donoghue et al., 2004c). Because binary logistic regression models group membership using a set of numerical variables, it has potential in performance analysis to investigate the effect of process indicator values on dependent match outcomes.

Discriminant function analysis is a further technique that predicts group membership using a set of numerical variables. The difference is that discriminant function analysis can be used to predict membership of one of more than two groups. Previous data with known group membership is used to produce a set of canonical discriminant functions that are used to plot new cases on a territorial map. This technique was used to predict whether group matches in international soccer tournaments would be won by the higher ranked team according to FIFA, be draws or upsets (O'Donoghue, 2005b, 2006c, O'Donoghue *et al.*, 2004c).

Factor analysis

Factor analysis is a data reduction technique that helps us reduce the large set of variables, where some pairs of variables are correlated, to a smaller set of independent factors. A correlation matrix shows the correlation between each pair of variables, assisting decisions about which to factorise. Each factor is a linear combination of all original variables, called an 'eigenvector'. Factors are extracted from the data starting with the one with the greatest importance in terms of how much of the variance in the original data it represents. The importance of each factor is represented by an eigenvalue, which is a numerical value of the number of original variables it is worth. The process initially produces the same number of factors as original variables, with some factors having eigenvalues greater than one and others having eigenvalues of less than one. The researcher needs to balance the need to extract a concise set of factors with the need to cover as much of variance in the original data as possible. Typically, 15 or more original variables can be transformed into a set of factors where more than 60 per cent of the variance in the data is represented by the first five or six factors. These five or six factors are extracted and must then be interpreted by examining the original variables that are highly correlated (loaded) to them. The factors can then be analysed in the same way as any other variables, except there are fewer factors meaning the analysis will be more concise. In performance analysis, factor analysis (using the principal components analysis technique) has been used to reduce a set of performance indicators to a more concise set of performance indicators that represent independent aspect of performance (O'Donoghue, 2008a).

Cluster analysis

Traditionally, research in sport and exercise has used methods based on the assumptions of the normative or the interpretive paradigm (Burrell and Morgan, 1979). The human nature assumptions of these paradigms motivate different approaches to data analysis. The normative paradigm assumes that human behaviour is deterministic and can be characterised by an average human being. The interpretive paradigm assumes that human behaviour is voluntaristic with individual responses to similar situations. There are occasions where neither of these approaches is satisfactory and it is better to identify different types of people. An example of this is how the strategy of tennis players is influenced by score line (O'Donoghue, 2003). There may be different types of player with respect to how strategy is influenced in different ways by score line states. For example, some players may apply a similar strategy irrespective of the score line while others may have score line-dependent strategies. Therefore, what is needed is a means of identifying different player types with respect to how score line affects strategy. These subgroups (or clusters) of players can then be analysed using the quantitative techniques. Cluster analysis examines the data in order to identify different clusters of player with respect to score line effect (O'Donoghue, 2003).

SUMMARY

In quantitative research, data analysis is very much dependent on research design and is dictated by the purpose of the study. It is often necessary to take sports performance data through several stages of processing before the variables to be analysed are created. It is very rare that students will be analysing a single sample without testing any relationships or differences. In single sample studies, the quantitative results will be restricted to descriptive statistics. The statistical tests used to test the strength of relationships between variables depend on the scales of measurement of the variables involved; relationships between nominal variables are tested using the chi-square test of independence whereas relationships between numerical variables use correlation techniques. When testing for differences between samples, the test to be used depends on the scale of measurement of the dependent variables, whether the samples are independent or related, the number of samples being compared and whether the data satisfy the assumptions of parametric statistical tests. The results of inferential statistical tests are the 'icing on the cake' but students should always include the all-important descriptive statistics as well when presenting their results. This is a quantitative data analysis chapter within a performance analysis textbook and students using specialist techniques are encouraged to also read statistics textbooks.

CHAPTER 9

ANALYSIS OF QUALITATIVE SPORTS PERFORMANCE DATA

INTRODUCTION

In writing this textbook, it was very important for the author to write this chapter on qualitative research without involving a second author. Qualitative research is clearly not an area of expertise of the author's, but there were very good reasons for wanting to personally write this chapter on qualitative research in performance analysis.

First, there are some who take dogmatic positions, not only with quantitative or qualitative research methodologies but also within particular quantitative approaches and particular qualitative approaches. As a lecturer of research methods, a dissertation supervisor as well as a lecturer of performance analysis, the author is keen that students demonstrate an appreciation of the relative strengths and weaknesses of quantitative and qualitative approaches.

Secondly, there may be some students who believe that qualitative research is an easy way out of doing statistics. It is very important that this author, a performance analyst, makes the case that qualitative research is very valuable, labour-intensive and requires the development and use of intellectual skills and knowledge by the student.

Chapter 2 has already described the underlying assumptions of qualitative methods and described the characteristics of research processes involving qualitative methods in performance analysis. This chapter compares qualitative and quantitative research in performance analysis, highlighting where the comparison between the two approaches is different in performance analysis to other areas of sports science. The chapter describes the nature of qualitative data in performance analysis investigations and how such data can be gathered using interviews, observational methods or accounts. The issues of authenticity, trustworthiness and validity of qualitative data, analysis of qualitative data, identifying and tracing themes and developing theory from such data are also covered. Finally, the chapter describes how findings from the analysis of qualitative data can be presented.

COMPARING QUALITATIVE AND QUANTITATIVE RESEARCH

Thomas *et al.* (2005: 346) compared quantitative and qualitative research in terms of use of hypotheses, the types of sample, the research setting, data gathering personnel, research design and data analysis. This is a useful starting point for comparing qualitative and quantitative research in performance analysis. There are similarities and differences to the contrast between these two approaches in performance analysis and other areas of sports science. The main differences between performance analysis research and in other sports science research relate to the setting, with some minor differences relating to samples and data gathering personnel.

The research setting

Much quantitative sports science research is done in laboratory settings (Thomas *et al.*, 2005: 346). Quantitative performance analysis research can be done in laboratory settings through biomechanical analysis of key techniques of the sport. There are other data that can be collected during laboratory analysis of the performance of such skills using EEG data for example (Loze *et al.*, 2001). However, there is a large amount of notational analysis research where sports performance is observed in its natural setting. This empirical observational research can be done live while watching matches at the venue or live on television, or post-match analysis can be done from match video recordings.

Samples

Samples in qualitative research are typically small and purposive (Flick, 2007: 28), while quantitative research aims to use large random samples (Thomas *et al.*, 2005: 346). Samples in notational analysis studies are indeed large, but some quantitative biomechanics investigations involve extensive testing and data processing such that samples need to be as small as in some qualitative research studies. The samples used in performance analysis are rarely random samples, but then this could be said of many other types of quantitative sports science research. The reasons why matches and players are not chosen at random include the use of television coverage where matches are selected by the broadcaster.

Data gathering

Thomas *et al.* (2005: 346) characterised quantitative research as being done by an objective measurement instrument. In performance analysis of sport, such an instrument could involve human operators. These human operators would typically follow objective measurement procedures so that data are independent of the personal view of the particular operators involved. However, there are systems where observational and data recording tasks are done under such time pressure that it is not possible to fully apply the stated criteria for classifying observed behaviour. A further consideration is that the operational definitions used in performance analysis of sport are often vague and it may not be possible to concisely present completely objective definitions. Where completely objective definitions are used, it may not be possible for human operators to apply them. For example, if running is defined as any forward movement at a speed of $5m.s^{-1}$ or faster, a human operator would not be able to judge where a player's movement exceeds this speed threshold to an accuracy of 0.1s. Such research should only be done by trained observers and the system must be tested for reliability. Therefore, the student often collects the data for their own research project, but will use at least one other trained observer for the purpose of demonstrating inter-operator reliability. Such a reliability test is also evidence that any data have been collected with a system that is reasonably independent of the author's subjective classification of behaviour even if the author collected all of the data for the main study.

There is performance analysis research where data gathering can be done by large teams of trained data collectors. The match statistics provided by IBM on the official websites of Grand Slam tennis tournaments will have been collected by trained system operators. The statistics for aces, double faults, first service points, points won on first and second serve are straightforward. However, the websites do not define what is meant by terms like 'winner', 'unforced error' and 'net point', and there are differences in what counts for each of these between different methods used in scientific research. For example, O'Donoghue (2007b) found that the data presented on the official tournament websites tended to give higher values for the percentage of points where players went to the net than would be the case if the author's own definition was used. It is, therefore, necessary to undertake a validity study to compare the results reported on different tournament websites with the results produced by other systems that have been demonstrated to be reliable. Any tendency for the websites to underestimate or overestimate any variables compared to other systems must be acknowledged by the researcher.

Use of hypotheses

The selection of a research topic in performance analysis involves a combination of both deductive and inductive reasoning. However, as in other areas of research, there will be a greater amount of deductive reasoning when developing hypotheses in quantitative studies than in qualitative studies. In quantitative research, induction from specific observations will not contribute as greatly to the development of research hypotheses as deductions from general theory, or from knowledge of coaching practice and media speculation.

Research design

Research design in performance analysis involves a similar contrast between qualitative and quantitative research as seen in other disciplines. Quantitative studies are designed before data collection commences and then follow a fixed design to gather and analyse data to test some research hypothesis. Indeed, some quantitative performance analysis investigations even use designs from previous investigations but with new up-to-date data or tailoring the methods to different sports. Even when formal hypotheses are not specified at the outset of the project, there is usually a defined purpose for the research in terms of vaguely specified variables and concepts to be specified during system development. There will still be a fixed design, the researchers will know that there will be a set of more objective performance variables of interest and the statistical procedures involved will also be known.

There is varying practice in qualitative research when it comes to the design of studies. A grounded theory is one developed from the data gathered without any preconceived theory being considered at the beginning of the research process (Strauss and Corbin, 1998: 12). However, grounded theory is just one of 16 perspectives of qualitative research described by Patton (2002: 131–5). Other types of qualitative research can be done to validate or extend existing theory and this involves deductive analysis. Indeed Strauss and Corbin (1998: 22), in their book on developing grounded theory, challenged the view that qualitative research never validates pre-existing theory.

Data analysis

The contrast in data analysis between qualitative and quantitative research is similar in performance analysis to the contrast described by Thomas *et al.* (2005: 346) between these types of research in physical activity in general. Quantitative data are analysed using statistical techniques while qualitative data are analysed using more interpretive methods.

QUALITATIVE DATA

Some of the methodologies that are associated with qualitative research are concerned with inner thoughts, attitudes, beliefs, emotions, intentions and experiences. It is not possible to study these in the form they exist within the human mind and so language is used to communicate these to others. Patton (2002: 478) recognised the role of language in qualitative data, stating that experience and perceptions were presented as readable narratives. Language provides abstract representations of thoughts, attitudes, beliefs, emotions, intentions and experiences. The data used in qualitative research takes the form of words that are used in descriptions of events, scenes, experiences and emotions to convey a mental image of them (Strauss and Corbin, 1998: 15). The mental image developed by the audience of the description may not be the same as that of the person communicating the description. This means that the quality of the data (the description) will largely depend on the ability of the person providing the data to use language effectively. Therefore, qualitative data should be rich word-based detailed descriptions and explanations. Interviews, field notes and documents are examples of sources of qualitative data that are used in research (Patton, 2002: 4–5). Qualitative data also include images (Silverman, 2005: 241–67), audio recordings and video recordings of events.

DATA GATHERING TECHNIQUES

Interviews

Interviews are used for a variety of different purposes including the investigation of crime, journalism, personnel recruitment and promotion, examination, medical diagnosis and therapy. Interviews, like questionnaires, are self-report methods rather than direct observation methods. The purpose of a research interview is to obtain qualitative data from an informant who is relevant to the research question of interest.

There are different purposes of using interviews within a performance analysis. In the early stages of a performance analysis study, an interview with an expert coach or a focus group of experts in the given sport could be used as a way of determining research questions, concepts, variables and hypotheses prior to system development and collecting sports performance data. Alternatively, once a quantitative performance analysis investigation has produced results, an interview with an expert coach in the given sport can be used to find explanations for the observed behaviour patterns. Interviews have the potential to be used in studies that aim to explain patterns of play in sports performance. These studies may not include any actual performance analysis data collection techniques, drawing only on performance analysis literature in developing the research problem to be addressed using qualitative research. For example, the author recently introduced Interacting Performances Theory (O'Donoghue, 2009a) based on purely quantitative evidence. The theory is composed of three main parts:

1. The process and outcome of a performance are influenced by the quality (based on World ranking for example) of the opposition.
2. The process and outcome of a performance are influenced by the type (based on the style of play, irrespective of success) of the opposition.
3. The process of a performance is influenced by the same opposition effects in different ways for different performers.

Having established quantitative evidence for all three parts of the theory, it would be useful to investigate the mechanisms that lead to the opposition effects observed. Why is the process (style of play) used by different performers influenced in different ways by the same opposition effects? Consider the example of women's singles tennis, where the percentage of points the opponent goes to the net affects the amount of points different players go to the net in different ways (O'Donoghue, 2009a). The percentage of points that some players go to the net is positively associated with the percentage of points that the opponent goes to the net. For some other players, the percentage of points where they go to the net is negatively associated with the percentage of points that the opponent goes to the net. There are also players who go to the net on a similar percentage of points irrespective of how often the opponent goes to the net. An interview study is a way of gaining essential evidence to explain why these different patterns are observed. The following questions might be central to such a study. Is this variation in effect of opposition behaviour caused by the perceived relative strengths and weaknesses of the players? Do anthropometric and physical limitations of players influence their style of play? Is the effect caused directly or indirectly by differences in preparation or genetic differences between players? Interacting Performances Theory is currently not a theory in the true sense of the word as no explanation for the observed patterns is supported by evidence. Performance analysis is an observational method that cannot provide this type of evidence and, therefore, other research methods such as interviewing are required.

Another example of an interview was in the study by Greene (2008) on 400m hurdles performance. Having devised performance indicators for 400m hurdles performance, Greene presented the results to an experienced international 400m hurdles coach who examined them prior to an open conversational interview about how such performance indicators could be used in elite level athletics by athletes, coaches and high performance directors.

There are different types of interview ranging from highly structured interviews to open interviews. The choice of interview type depends on the number and type of participants, the purpose of the interview study, whether or not there is an initial theoretical basis to the study as well as the breadth and depth of data required for analysis (Patton, 2002: 341–8). The number of participants depends on the purpose of the study: investigations with solely interview data involve about 15 participants while biographical interviews will involve a single participant (Kvale and Brinkmann, 2009: 113–15). Phenomenology interviews are used to investigate the lived experiences of the participants whereas ethnographic interviews are used to investigate the culture of a group of interest (Marshall and Rossman, 1999: 104–5). Marshall and Rossman (1999: 105–6) described 'elite' interviewing as an interview for investigating prominent experts in a given field. This is relevant in sports performance research to uncover processes, activity and support structures that lead to successful performance. There are disadvantages to interviewing elite performers and/or elite coaches, which include their limited availability and the fact that competitive sport often requires successful participants to guard the secrets of their success. It is very important that researchers doing elite interviews have the knowledge and experience to be able to conduct such an interview, recognising important areas and probing into them (Kvale and Brinkmann, 2009: 147).

A variation of interviewing is the use of a focus group, which brings together a small group of similar people to discuss a specific topic. This can be very useful where a squad has been working together and the views of different members are important. People of different opinions can be brought together and the discussion can yield many useful research findings. Focus groups provide quality controls as participants tend to provide checks on each other's input. A disadvantage of focus groups stems from the dynamics of the communication involved, which does not always permit probing questions to follow up a particular point. Other disadvantages include formal or informal power positions within the group that make some members reluctant to state views.

An interview study involves a preparation phase, a data gathering phase and a data analysis phase with the last two phases possibly overlapping. In some studies all the interview data are collected before data analysis commences (Holt and Mitchell, 2006) while in other studies interviewing and data analysis can overlap (Côté et al., 1995). During the preparation phase, the researcher does background research into the subject area of the research to identify the main purpose of the research, sub-topics of interest and the specific information needs required from the interview study. The extent of this background work and the initial structure of sub-topics and information needs depend on where the study lies on a continuum of exploratory grounded-type studies to studies that are testing some pre-existing theory. The scope of the study should be set and criteria for the inclusion of participants should be established. An interview guide can be produced including basic themes, but this should be a guide to the interview rather than a restrictive control mechanism (Patton, 2002: 343–4). Davies et al. (2005)

provided an excellent example of an interview guide as an appendix to an interview study on job satisfaction of basketball coaches. This was a four-part interview guide that had sections for background information, core questions, summary questions and additional information from the participants. Each section consisted of one to three open questions with up to four areas for potential probing for each question listed in the interview guide. The student may wish to take notes to remind themselves of other areas to be raised later in the interview. The student should explain before the interview commences the nature of any note taking that might occur during the interview.

During the interview, the student should maintain a rapport with the participant and use open ended questions to allow the participant to discuss the issues that are important to them. Kvale and Brinkmann (2009: 134–8) described different types of questions to be used within interview studies. These include introductory questions, follow-up questions and probing questions. Introductory questions are used to allow the informant to talk about an area of interest, while follow-up questions encourage the inform- ant to elaborate on these areas. Probing allows more detailed information to be elicited, increasing the richness and depth of the data that is gathered (Patton, 2002: 372–4).

Before detailed analysis of interview data can occur, it is necessary to establish the trustworthiness of the data. Trustworthiness means that the data provided represent the participant's views and that the participant has been portrayed fairly. The trustworthiness of the analysis should also be established. This is described later in the current chapter.

The analysis of interview data and the analysis of other qualitative data have some differences due to the interview and account data coming directly from the participants, while observational data are descriptions made by the researcher. Those students who are doing in-depth interviews as the main data collection activity of their research project should read specialist texts on interviewing (Kvale and Brinkmann, 2009) and on qualitative data anal- ysis texts (Miles and Huberman, 1994, Patton, 2002, Silverman 1993, 2005, Strauss and Corbin, 1998). A section on qualitative data analysis appears later in this chapter.

Observation

Laird and Waters (2008) have shown that qualified soccer coaches can accurately recall an average of 59 per cent of critical events that occur during a soccer match. This is a higher value than the 42 per cent reported by Franks et al. (1983), but is still not a perfect recollection of the game. Human observation during field research can suffer from a similar problem due to highly selective human perception and subconscious bias (Patton, 2002: 264). It is, therefore, necessary for students intending to do field work to train and develop observational skills. Patton (2002: 260–61) listed six things that need to be included in observer training:

1. Learning to pay attention (seeing and hearing important detail).
2. Practise in writing descriptively.
3. Disciplined recording of field notes.
4. Being able to distinguish between important details and less important background data.
5. Using rigorous methods to validate and triangulate observations.
6. Acknowledging the researcher's own strengths and limitations.

The advantages of direct personal contact with the setting of the observations are that the context of people's actions and interactions is understood without having to rely on preconceptions, the trained observer will note things that participants would take for granted, and the observer will see things that participants might be reluctant to discuss (Patton, 2002: 262–3). There are different forms of observational research that can be classified by whether the researcher participates in the behaviour or is an onlooker, whether the observation is done covertly or overtly, whether there is a single researcher or a team of researchers, the duration of the field study, and whether the observations have a broad or narrow focus (Patton, 2002: 265–76).

Once an observational field study is planned and approved, there are various stages to field work which commence with entry into the field. This may involve negotiating entry with gatekeepers (Patton, 2002: 310–26). In covert observational research, the researcher would need some role as cover. This involves deception and is usually only approved by university ethics committees for the most experienced researchers doing critically important research that cannot be done any other way. Once the researcher has established a role in the field, they must build trust and maintain relations in the field and adjust to the setting. The researcher should try to be as unobtrusive as possible to minimise participants altering behaviour due to the knowledge that they are being observed. The researcher must remain flexible and record all important details, including those that do not appear to relate to themes anticipated at the beginning of the study. Observations are done during planned activities of the group under observation but also during unplanned activity, which might provide valuable data. The use of language is also important and while it may not be possible to record precise quotes, the researcher should try to record as much detail as possible of important statements made. In various work settings, participants tend to develop their own dialect including key technical terms that are understood by group members but may be meaningless to those outside the group. For example, Sugden (2002: 21) described 'tout-speak' as a collection of terms used by ticket touts. This is an important part of group culture that should be recorded and described. Non-verbal communication is also important and the way in which it is used should also be described as much as possible.

Once the reporting deadline approaches or the data become theoretically saturated, the researcher should aim for closure by using final field contacts to verify aspects of emerging hypotheses. Patton (2002: 324–6) described an evaluation stage where researchers provide formative feedback to participants that may be very useful for the participants' practice as well as acting as a verification process for the research. This process is ideal for a researcher working as a performance analyst in order to experience the role within a squad. There may be ample opportunities to write up detailed field notes in between match recording and briefing sessions.

The researcher records data about observations in field notes that contain descriptions of the physical settings (Silverman: 2005, 17–19), the social environment, events (Flick, 2007: 30-31), complex actions, interactions, behaviours, and decisions and how these are communicated. Artefacts, records and archives can also be examined and recorded. When recording field notes, the researcher should remember that the reader will not have seen the setting and will be relying on the portrayal made by the researcher. Therefore, vague descriptions should be avoided. Field notes are essential to later analysis and should be detailed non-judgemental descriptions (Marshall and Rossman, 1999: 98).

Field notes are unstructured descriptions of events, actions and thoughts that are generally unplanned (Gibbs, 2007: 27). Field notes should be written up as soon as possible during or after field contacts. Spradley (1979: 74–6) described the use of four levels of field notes. Short field notes are recorded as soon as possible during or after field contacts. Expanded field notes are written after this and include more detail than it would have been possible to record when writing short field notes. A journal can be kept, recording problems that occur during field work. The final level of field notes is a provisional running record of analysis. Gibbs (2007: 26) describes the use of a research diary in which ideas and comments about discussions with other researchers can be written.

The researcher's own behaviour, how they are being treated (Silverman, 2005: 158), feelings, reactions and reflections should also be recorded within field notes as well as initial interpretations and working hypotheses. The researcher may have inspirations during field work that should also be recorded along with reminders of areas for use in subsequent field contacts (Patton, 2002: 302–5). Initially, a checklist of codes is not used, but as more and more field contacts are experienced, field work becomes more focused, codes can be developed and a context-sensitive checklist can be used (Marshall and Rossman, 1999: 98–9) with field notes becoming more theoretically focused as research progresses (Silverman, 2005: 177–80).

Accounts

Those who participate in a behaviour can provide accounts that describe their motives and beliefs in relation to the behaviour. Cohen et al. (2007: 384–95) described a process by which participants could be selected to write accounts and then the accounts could be written by them before being gathered and analysed by the researcher. However, this does require considerable effort from the participants that they might not be willing to expend for the purpose of the student's research project. Alternatively, there are pre-existing accounts in the form of autobiographies, publicly available video recorded interviews and documentaries. The material presented within these sources is not necessarily in a format that the student would have chosen. Nevertheless, such sources still have value within research studies. An excellent example of the use of published accounts is Brown's (2005) study comparing the race strategies of those female track athletes who attempted the 800m and 1,500m double during the 2004 Olympic Games with that of those athletes who only entered the 800m or only entered the 1,500m. This study used a triangulation approach involving an analysis of split times and positioning during races, an interview with a former Olympic track champion who had attempted a double, and an analysis of published accounts by one of the athletes who attempted the 800m/1,500m double in the 2004 Olympic Games. These accounts included a published diary specifically covering the preparation for and competing in the 2004 Olympic Games and an autobiography (Holmes, 2005, Lewis, 2004). The qualitative analysis of these accounts and the interview with a previous Olympic champion allowed aspects of attempting the 800m and 1,500m double to be investigated that could not be investigated using observational methods alone. These aspects were the main themes of the qualitative analysis:

- preparation for an Olympic double;
- influences on decision to attempt the double;
- the pressures of attempting the double;
- racing in two Olympic events;
- tactics when attempting an Olympic double.

ANALYSIS OF QUALITATIVE DATA

A basic qualitative analysis technique

The use of qualitative methods may be a small or large part of a performance analysis research project. Some of the examples used in this chapter have been single interviews to help explain patterns seen in quantitative data (Brown, 2005, Donnelly and O'Donoghue, 2008, Greene, 2008). Applying some of the data analysis techniques and associated software packages used in psychology research (Côté et al., 1995) to such small-scale qualitative analysis would be like using an elephant gun to catch a butterfly. These methods use layers of components, categories and properties to describe the structure within the data. Therefore, some students have used a much simpler approach to qualitative data analysis. The process typically commences with verbatim interview transcripts being produced, although the simple analysis technique described here is also suitable with detailed descriptions within field notes. The student may have a preliminary set of three to six broad themes that they wish to explore within the interview data. The student reads the transcripts while listening to the audio recording of the interviews to decide if any themes need to merge or if any new themes are required.

The next stage of the process is to analyse the first interview with respect to the themes. The student uses a different colour of highlighter pen to represent each theme (these highlighter pens typically come in turquoise, green, yellow, pink and orange so five themes is ideal for this types of analysis). Anywhere where the interview contains evidence that is relevant to a particular theme, the student applies the appropriate colour of highlighter pen to the transcript at that point. There are occasions where a sentence relates to more than one theme and so two different colours are used to highlight this evidence. Despite not being modern technology, pen and paper have advantages over the use of software packages for analysing qualitative data. The student does not have to spend time training to use a software package for a short-term research project. The student will be able to write additional notes, reminders and comments in the margins of interview transcripts and on the back of the sheets of paper the transcripts are written on. Other annotations can be made, linking related responses within interview transcripts.

This first interview should then be analysed by an independent researcher using a new printout of the transcript. The independent researcher will be advised of the themes represented by each highlighter pen colour and how the text is to be highlighted where there is evidence of one or more themes. Once this is done, the student can compare the two highlighted transcripts, noting the sentences associated with each theme where the independent analyses agreed and also any disagreements that occurred. There are two main types of disagreement: first, one researcher may associate a sentence with a theme and the other researcher may not highlight the sentence at all; secondly, both researchers may highlight a sentence but associate it with different themes. Disagreements are often not complete black and white disagreements and the student should report the types of sentences that were interpreted as being relevant to different themes. This gives the reader a better idea of reliability of the analysis than a simple percentage error of sentences where the researchers disagreed about the theme represented.

Once this reliability study is completed, the student should complete the highlighting of the remaining interview transcripts. Once this is completed, the student is able to draw some conclusions about each interview with respect to each theme. This should be done on a separate summary analysis form for each interview. Once this is completed the student analyses the data for each theme separately across the full set of interviews. This is done by reading only the evidence relating to a given theme at a time. For example, if a particular theme is highlighted in green, then the student reads all of the text highlighted in green from all of the transcripts to draw some general conclusions for that theme. These general conclusions are then checked against each interview transcript using a process of negative case selection. This involves the student reading through all of the evidence relating to the theme from all of the interview transcripts in an attempt to try and disprove their own conclusions. Where any discrepant evidence is found the student should revise the conclusions so they are consistent with the data in the interview transcripts. Once this process of analysing a theme in isolation has been completed for each theme, the student is ready to report the results of the qualitative analysis. This simple approach using highlighter pens can also be done with notes made from qualitative observation of video recordings of behaviour (Paisey and O'Donoghue, 2008). It must be stated, though, that this type of analysis requires detailed descriptions of observations that are of the quality one would expect from field notes.

Rigorous qualitative analysis

The qualitative analysis process explained in the previous section has been sufficient for a number of performance analysis research projects where interviews have been done to complement the use of quantitative performance analysis methods. However, if students are aiming for high marks or the publication of their study, more ambitious qualitative analysis techniques should be used. A further consideration is the proportion of the data collection and analysis that is done using qualitative techniques. In some dissertations, this may be more than half of the data collection effort, warranting a more comprehensive analysis of qualitative data using approaches that have been developed and used by leading scientists in published sports science research.

There are many different ways of analysing qualitative data and these have been described in specialist texts on qualitative research methods (Marshall and Rossman, 1999, Miles and Huberman, 1994, Patton, 2002, Silverman, 1993, 2005, Strauss and Corbin, 1998). Prior to writing this chapter, the author read these and other materials with the intention of recommending an approach suitable for use in performance analysis projects. All of the approaches have their strengths and weaknesses and so it would not be appropriate to exclude particular approaches to the analysis of qualitative data. There is, however, an approach described here that the author found suitable for many types of qualitative research that could be done within performance analysis projects. This is the approach used by Côté *et al.* (1995) to analyse expert gymnastics coaches' knowledge. The reasons why this particular approach appealed to the author of this book were, a) the method was used to study coaches' knowledge and performance analysis is often used within a coaching context, b) the method is a systematic way of tracing how evidence supports theory, c) the method is adaptable and can be used in an inductive or a deductive manner, and d) Côté *et al.*'s (1995) paper was published in the *Journal of Sport and Exercise Psychology* and the analysis of psychological aspects of performance is a main motivation of using interviews in conjunction with performance analysis methods.

Côté *et al.*'s (1995) method was used in a bottom-up inductive manner to develop a grounded theory of expert gymnastics coaches' knowledge. In this method, interview transcripts (or other unstructured qualitative data) are read in detail to identify meaning units (MUs), which are meaningful pieces of text that can be used as evidence supporting an emerging theory. Figure 9.1 shows that similar meaning units are then grouped into properties that are then grouped into broader categories and finally into the main components of the theory. This inductive content analysis moves from the evidence in the interview transcripts, gradually raising the level of abstraction. The hierarchy of components, categories, properties and meaning units adds structure to the data that is useful when interpreting the meaning of the data. However, this structure should guide the analysis rather than replace the raw uncategorised data. One of the strengths of qualitative data is that they are rich in descriptions made by the participants (or the researcher in the case of field notes) and the quality of the data should not be bypassed during the analysis.

Figure 9.1 shows that this approach of using different levels can also be applied in a top-down deductive process where the purpose of the qualitative analysis is to validate a theory or where themes have been anticipated at the beginning of the analysis. The MUs will still be separated from less

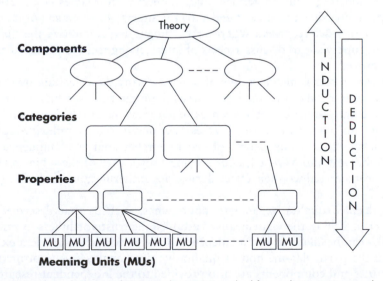

Figure 9.1 Adapted version of Côté *et al.*'s (1995) method for analysing unstructured qualitative data

important text within the interview transcripts during the initial stages of the process. However, the MUs will initially be associated with broad themes (components) of the theory being validated before more detailed subcategories and properties are introduced when analysing specific manifestations of the theory within the data.

There are software packages that can be used during qualitative analysis such as the NUD-IST (Non-numerical Unstructured Data with Indexing, Searching and Theorising) package (QSR International, Doncaster, Victoria, Australia). These systems are flexible and have an additional advantage of being standard, which means that they can be used by independent researchers within reliability studies more easily than non-standard methods. Readers interested in the detail of features provided by these software packages are directed to the coverage of such system by Gibbs (2007: 105–42).

Trustworthiness of data

Trustworthiness is a quality of interview data that is concerned with the accuracy with which the research portrays the attitudes, motives and knowledge of the participants. There are academic journals where methods sections of most papers describing interview research include a specific subsection on trustworthiness. There are a number of steps that can be taken to establish the trustworthiness of interview data. The interviewer's knowledge of the subject area has been reported as contributing to trustworthiness in some research (Salmela et al., 2006). Researchers should develop interviewing skills by first reading texts on effective research interviewing and then conducting a pilot interview under supervision (Davies et al., 2005). The pilot interview could be attended by the supervisor in person or it could be audio or video recorded. Where a recording is made, it allows the student and the supervisor to discuss aspects of interviewing technique that need to be improved.

Observer bias is also a source of untrustworthy data; leading questions should be avoided during the interview as these can be based on interviewer bias and can influence the responses made (Patton, 2002: 367). Another way of reducing the effects of any researcher bias is to use a reflexive journal that acknowledges any beliefs of the researcher that may influence the research (Holt and Mitchell, 2006). Being fully aware of these biases helps the researcher reduce their effect during the gathering and analysis of the data.

In detailed qualitative analysis approaches, such as that described by Côté et al. (1995), there are means of establishing trustworthiness. A sample of MUs can be analysed by an independent researcher with sufficient expertise in the particular method of qualitative data analysis. The properties, categories and components are also provided to the independent researcher who will have been trained in the content analysis scheme (Horton et al., 2005) and who has the task of matching the MUs to these properties, categories and components (Davies *et al.*, 2005). Any discrepancies between the matching made by the student and that of the independent researcher can be discussed and issues clarified (Davies *et al.*, 2005).

Member checking can be done to ensure that the interviewees are satisfied that they have been portrayed fairly (Holt and Mitchell, 2006, Salmela *et al.*, 2006). Interview transcripts can be provided to the interviewees prior to analysis for accuracy checks. The results and conclusions can be provided to the interviewees after the analysis is done, requesting feedback on any necessary rephrasing, information to be omitted or additional information that should be included.

Reporting the findings of qualitative analysis

If using the basic method of highlighting evidence relating to themes with coloured highlighter pens, the findings of the qualitative research can be reported using a subsection of the results chapter for each theme. Having read all of the interview transcripts (or field notes) relating to a given theme, the student should write a paragraph describing what has been found about that theme. A major difference between the analysis of qualitative data and the analysis of quantitative data is that the findings presented after qualitative analysis can include exerts of raw unedited data. There may be short or long passages of text in the qualitative data that characterise the theme. This is particularly useful with interview data or accounts because the words will be those of the participants, allowing the reader to make their own interpretations of the data. Where an inductive approach is used to develop a theory, the results can describe the components of a theory and their interrelationships using a diagram; a good example of this is the coaching model (Côté *et al.*, 1995). The individual components of the model can then be explained using excerpts of raw data to reinforce the findings. More detail of writing up qualitative research is found in Chapter 10.

SUMMARY

Quantitative observational techniques are good for describing 'what' participants do during a performance, but they are not so good at explaining 'why'. Therefore, qualitative techniques have great potential in the analysis of sports performance. In this chapter the main differences between quantitative and qualitative research have been explained with reference to how these techniques are used in performance analysis research. Theory can be developed inductively (grounded theory) or deductively (to validate an initially speculated theory). The main qualitative data gathering methods are interviewing, non-participant observation and accounts. Data from these can be analysed using similar methods, although some differences are necessitated by the fact that interview and account data come from the participants while observational field notes are the words of the researcher. Member checking (by the participant) and independent analysis of qualitative data using the themes or category structure devised for the research helps to establish the level of 'trustworthiness' of the data, its analysis and the theory supported by it. The results summarise the findings of the analysis in the form of abstract portrayals of general patterns, diagrammatic models and supporting excerpts of raw qualitative data.

CHAPTER 10

WRITING UP PERFORMANCE ANALYSIS RESEARCH

INTRODUCTION

This chapter outlines how performance analysis research should be reported, with the structure of the introduction, literature review, methods, results, discussion and conclusion chapters being covered. The description of qualitative methods, reporting the findings of qualitative analysis and their interpretations are also covered in this chapter. Different types of research report are discussed: research papers, conference presentations, posters, research proposals and theses. All research projects are different and there will be varying practice in how research is written up depending on the audience of the report, and requirements and constraints specified for the research report. The previous chapters of this book have already covered some details of how research is to be written up. For example, Chapter 4 showed how formal hypotheses are to be presented, while Chapter 6 covered the contents of the methods chapter of a dissertation. Therefore, there may be some necessary overlap between this chapter and previous ones, with the main focus of the current chapter being the reporting of the research.

RESEARCH PROPOSALS

Student research proposals

Research proposals are made by students at both undergraduate and post-graduate levels for assessment and approval purposes. Two key elements of the assessment are the academic and ethical aspects of the proposed research. The academic aspects include the relevance of the project to the student's programme of study, the volume of work being proposed, the academic level of the work, practical issues involved in the research and the academic interest of the proposed research. Chapter 3 covered the criteria for selecting a research topic, but the extent to which these criteria have been satisfied must still be checked by those assessing the research proposal. The student is doing a degree programme that leads to a recognised and valuable qualification. Therefore, the research project must be relevant to the degree programme. The proposed research may involve less data gathering and analysis than a typical performance analysis coursework and will, therefore, need to be made ambitious enough to have a chance of being successful as a research project. Alternatively, the proposed research may involve a volume of work that is so high that other elements of the student's final year study will suffer. When considering the academic aspects of the project, the assessors could give a pass grade to the research proposal as part of a research methods assessment but not approve the proposed project. There may be practical or logistical reasons for not approving well written proposals. Alternatively, there may be research proposals that are awarded a fail grade as part of the assessment of a research methods module, but where the proposed research is approved as a research project to be done by the student. This might happen when the assessors recognise that the proposed research involves a sufficient volume of work of the required academic level but the proposal has been written up poorly, failing to justify the research question, failing to critically review relevant background research or failing to plan the research activity satisfactorily.

The other key element that is considered during the assessment of research proposals is ethical issues. This has been covered in Chapter 5, but it is worth mentioning here that projects that may be approved on ethical grounds might not be approved on academic grounds. For example, proposed research involving the analysis of public domain broadcast coverage of a sport may be granted ethical approval but may fail to satisfy the criteria for a research project. It is also possible that a research proposal may be awarded high marks as part of the assessment of research methods but may be rejected on ethical grounds. A research proposal that is rejected on ethical grounds cannot be approved as a research project.

The assessment of student research proposals is typically done by members of staff during the assessment of research methods modules or during the early stages of the research project module. The precise mechanism for dealing with research proposals depends on a number of factors including the size of the sports studies department and the volume of the proposals to be considered. Thomas and Nelson (1996: 389–91) described proposal committees that approve and reject proposals, requiring some students to appear in person to answer questions posed by the proposal committee. Many universities have brought the assessment of undergraduate research projects forward to before the final year of the programme of study to allow assessment, ethical consideration and any amendments and reassessment to be completed so that the research project can commence early in the student's final year.

A research proposal broadly contains the background, definition, scope, methods and logistic issues for the proposed research project. The introduction should critically review relevant literature, demonstrating a need for the proposed research project. The definition of the research project contains its aims, purpose and any hypotheses that are specified. Methods should be written in sufficient detail to allow the proposal to be considered on ethical grounds. Unlike many other research reports, the methods sections of research proposals are written in the future tense rather than the past tense. Kumar (2005: 188–9) elaborated on the material to be included within a research proposal, listing the following contents:

Introduction
Theoretical framework
Conceptual framework
Objectives
Hypotheses
Design
Methods
Any ethical issues and how these will be managed
Planned chapters of the eventual research report
Limitations
Timeframe

The theoretical framework is the background theoretical knowledge that underpins the study. The conceptual framework is developed from the theoretical framework and is the basis of the particular research project that has been proposed. Therefore, the conceptual framework has a narrower focus than the theoretical framework.

Many students include a schedule of research activity, which is a very useful mechanism for planning, management and control of the research project when it is being undertaken. However, students should also include a plan of their eventual research report as it will help the student identify write-up tasks that need to be included in the schedule of work. The research proposal is not merely a means of assessing research methods, but when a proposal is approved it acts as a contract between the university and the student (Thomas *et al.*, 2005: 371). The research to be done within a research project must only include the activity that has been approved by the university. Some research proposals are necessarily vague about the methods that will be used, requiring the flexibility for methods to change as the research project progresses. Such projects can be still approved as long as conditions for the research are set. This is also a problem for performance analysis research projects where work within the project will include the development of an analysis system. To prevent too much project activity being done in the proposal rather than in the eventual research project, it is often necessary for performance analysis projects to be vague about the performance indicators to be used and how they are to be gathered. It is still possible for such proposals to be approved by specifying strict conditions about the sources of data, for example. In writing the research proposal, students may consult with potential supervisors about their ideas but also about the structure and content of the research proposal. If the proposal contains ethical concerns that will need to be considered by research ethics committees, it is particularly important that the proposal is written well enough to allow the committee to make a decision. There is a balance that needs to be made in a member of staff's contribution to the research proposal. The research proposal is often an assessed piece of work that is contributing to the student's degree and, therefore, while the supervisor should give advice to help ensure the project will be of a satisfactory standard, the proposal must ultimately be the student's work.

Research grant applications

In addition to research proposals being done by students, academic staff at universities also write research proposals. Sometimes these proposals are applications for research funding on which the study depends. The source of research funding could be an internal research budget within the university or an external funding body. The funding body will have guidelines to be followed within the research proposal and often will not tolerate any deviation from these guidelines whatsoever. Research proposals being submitted as applications for research funding have additional contents above those of the research proposals already discussed. In particular, it is essential to provide a full budget for the research project, including staff, equipment, consumable and travel costs, and any other expenses that need to be paid from the research budget. The funding body will need to know about the people and organisations undertaking the research in order to make an assessment of their ability to undertake the proposed work. In evaluating the research proposal, the funding body will not only consider the theoretical and practical value of the research, but will also consider the cost of the research project.

There are other applications for research projects within priority areas specified by funding bodies. For example, sports governing bodies or local government may advertise for an independent organisation to evaluate some scheme they have been running. While the overall purpose of the project and the money available is set by the funding body, there are still extensive details that must be provided by applicants for the research contract related to the specific way in which the project will be carried out, the staff involved and their capabilities.

CHAPTERS OF A THESIS

The contribution of the research project to a university degree programme will vary from university to university, but typically it will be worth about one third of the final year of the programme of study. In the author's own university, the value of the dissertation has recently reduced from 48 UK credit points to 40 UK credit points out of total of 120 UK credit points for the final year of the degree programme. The contribution of the research project to the overall degree will dictate the required effort hours involved and the word limits that may apply. Undergraduate research projects have word limits of around 10,000 words (a minimum of 9,000 words and a maximum of 10,000 words at the author's university) for the main chapters of the dissertation. Dissertations within taught Master's programmes usually have word limits around 15,000 words. It is possible that word limits may also be specified for particular chapters, but usually they are not. Therefore, students need to set word targets for the different chapters based on marking schemes that have been outlined for dissertations and the emphasis of their own research project. Performance analysis projects often involve the development and testing of a performance analysis system as well as the use of that system within a research project into some area of sports performance. Therefore, the methods chapter would include more words than methods chapters in some other disciplines. The following list is a guide for the volume of words to be included in the different chapters of undergraduate performance analysis dissertations:

Introduction: 500–800 words.
Literature review: 2,000–2,500 words.
Methods: 1,500–2,000 words.
Results: 1,000–2,000 words; remember tables and figures may not contribute to the word limits specified.
Discussion: 2,500–3,500 words.
Conclusions and recommendations: about 400 words.

The number of words to be included within each chapter is largely influenced by the emphasis of the research project and where the major challenges of the project exist. Students should begin writing chapters as early as possible and should keep a notebook to record material to be included in the thesis that they might otherwise forget. Due to the undergraduate research project being done over a period of six to eight months, students should keep backup copies of their data as well as the chapters they are writing. This should be straightforward to do at least once a week with the reduced costs of CDs, memory sticks and external hard disks. Some e-mail providers allow sufficient space within e-mail accounts to permit students to e-mail their work to themselves so that it is also stored as attachments. The contents of the particular chapters are briefly outlined in the following subsections of the current chapter.

Introduction

The introduction of a research project includes a brief background to the area, the purpose of the study, a justification for the study, definition of key terms and any formal hypotheses in terms of these, the scope of the study and any limitations that the reader should be aware of at the outset of the dissertation. The background is not a full literature review but a brief outline of the research topic that generates an interest in the research problem. The justification for the study is a rationale for undertaking the study that explains why it was important for the study to be done and potential uses of the results. There is varying practice for when the introduction is written, with some supervisors wanting the introduction chapter to be completed before any other work commences while other supervisors are happy for the introduction to be done in a draft form early during the project before being completed after the literature review, methods, results and discussion have been completed. This is because the introduction is introducing the reader to a project that has now been done and there may be important details that need to be included in the introduction that could not have been known early in the research project.

Literature review

The review of literature is often done while selecting a research topic of interest and determining a particular research question. However, material does not have to be presented in the order it was written. The process of reading and reviewing literature has been covered in Chapter 3 and so the current section is concerned with how the literature review chapter should be structured and presented.

After the research project has been justified and specified in the introduction chapter, the role of the review of literature chapter is to provide a critical coverage of previous research as well as theoretical and methodological issues that are relevant to the research topic. The review of literature needs to be presented in a logical structure, so students should carefully consider the subsections that will be included within the review. A common mistake that students make is using dissertations from several years earlier as models of how to write up. There was a time when notational analysis was a new area of research and the use of computerised systems within notational analysis research was novel. Therefore, a review of notational analysis, its history and the benefits of using computerised systems may have been acceptable several years ago. However, such areas are far too general to be included within the literature reviews being done by undergraduate students today. Including such general material within the review of literature chapter is like a sports psychology (or biomechanics or physiology) project student having an 800-word subsection of their literature review introducing the discipline of sports psychology (or biomechanics or physiology).

There are different approaches to structuring the literature review. Some very good review of literature chapters commence by covering broad theoretical debates before introducing sport-specific evidence relating to the theory and eventually funnelling down to that literature which is most related to previous performance analysis research into the specific area of performance in the specific sport. Other very good literature review chapters may choose to use subsections for different components of the research problem before bringing them together at the end to identify a gap in the literature to be addressed by the student's research. Chapter 1 of this book identified 11 different broad research topics in performance analysis, some of which may overlap within a student's research project. For example, a student might be undertaking research into work-rate of referees or the tactics of players who have different levels of technical effectiveness. Therefore, it is impossible to propose effective literature review structures for all types of projects and so the four examples given here are included to give an understanding of the way in which literature reviews can be structured.

The first example is dissertations on technical effectiveness of different samples of players. There will be some independent factor that performance indicators measuring technical effectiveness are hypothesised to depend on. The samples could be independent (for example positional role or level of play), or related (for example period of the match, score line or venue) and the sport could be an individual sport or a team sport. Such research projects could decompose the literature review into two broad subsections. First, the performance indicators typically used to measure technical effectiveness in the particular sport could be identified within a critical review of previous research. This evidence would not only include performance analysis literature but also coaching literature that supports the analysis of particular techniques. Secondly, previous research into the effect of the chosen independent factor on technical effectiveness in the sport being investigated by the student, as well as in related sports, could be included.

The second example is dissertations on the influence of some independent factor on work-rate in a chosen sport. These types of dissertations could use three broad subsections. The first could be a critical review of literature into intermittent high intensity activity (or other types of activity depending on the sport of interest), which would include evidence from laboratory-based studies, field studies and observational studies. The second subsection could be a review of research evidence into the influence of the particular factor that is hypothesised to influence work-rate. This factor could be positional role, level of play, age group, gender, score line, period of the match or one of many others. As much evidence as possible relating to this factor should be drawn from literature in the specific sport of interest. However, this might be limited, so literature covering the effect of the factor in related sports may also be included. The third subsection of the literature review is the coverage of methods of analysing work-rate, their strengths and weaknesses.

The third example is dissertations on the effectiveness of instructional feedback to support for coaches and athletes. Such research projects could structure the literature review chapter into three subsections. The first could be a review of theory relating to feedback and skill acquisition. This can bring in coaching literature as well as motor control research. Where the instructional feedback relates to tactical aspects of performance, this first subsection would be better reviewing coaching literature into decision making and possibly sports psychology literature into attentional focus, anticipation and decision making. The second subsection of the literature review could be a critical review of different types of feedback and particular performance indicators that could be used in instructional feedback. These can be identified from purely observational studies as well as studies and professional reports examining the use of such indicators. The third subsection could be a critical review of previous research into the effectiveness of performance analysis-based feedback in sport.

The fourth example is dissertations on the behaviour of different samples of coaches based on some factor such as the age group of the players being coached. Such projects could include two large subsections. The first would cover theory of coaching style and how it is influenced by the hypothesised factor of interest. This literature could include coaching science material, sports psychology papers and performance analysis research. This first subsection could be further decomposed, but this might cause overlap and repetition between the smaller subsections. The second large subsection would be a critical review of different methods of measuring coach behaviour and the variables involved.

Methods

There is a mistake that is made in the majority of performance analysis dissertations read by the author, which is referring to a methods chapter as 'methodology'. Methodology is the discussion of alternative methods and approaches that can be used, their underlying assumptions, strengths and weaknesses. There are dissertations that do precede the description of methods with a justification for the methods used, and such chapters are rightly referred to as 'methodology chapters'. In performance analysis, the chapter that many students refer to as a 'methodology chapter' simply contains a description of the methods used in the student's research project. Performance analysis students typically describe their methods, having already discussed alternative approaches within the literature review chapter.

Where a methodology chapter is done, the contents would typically include a first section that is a justification for the methods used and a final section that evaluates the methods used (Gratton and Jones, 2004: 234). The sections in between these sections would be the same sections that would make up a methods chapter without the justification and evaluation of methods. Where a typical notational analysis study is being written up, one would expect to see sections on research design, system development, reliability evaluation, matches (or participants, or performances), data gathering and data analysis. As has already been mentioned in Chapter 6, the section on system development should include descriptions of the performance indicators being used, pilot work during system development and the final system, especially action variables to be entered and operating procedures. This information is vitally important to allow the study to be replicable, and even if precise operational definitions cannot be specified, the student should describe the different variables in enough detail to allow the reader to understand them.

Some performance analysis dissertations have included reliability results within the results chapter of the dissertation. This approach is criticised because the results of the reliability study may be required to make decisions about which performance indicators can be used and how they should be analysed. For example, a performance indicator may be unreliable when specified precisely, but if a categorical version of the performance indicator is used based on sub-ranges then the results may be reliable. This decision requires the reliability results and it is not desirable for the reader to have to go from reading the methods to reading the results to going back to reading the methods again. Another reason for including the reliability results in the methods chapter is that these results are not the results of the study but the outcome of a test of a system used to undertake the study. There are, of course, some studies that have the sole purpose of investigating the reliability of a system or alternative systems. Such studies should place the reliability results in the results chapter because they are the results of the study.

The section on data gathering should describe all action used to collect data including any filming or video recording of matches. The details of the matches (or performances) analysed in the study should include the level of play, criteria for inclusion of matches and the season(s) when the matches were played. The data analysis section should describe the processing of system outputs to produce the performance indicators being tested and the different statistical tests that were used.

There are dissertations in the wider area of performance analysis of sport that can be done using solely qualitative methods. For example, performance analysis support to a squad of athletes over a period of four months is a valuable experience that could be investigated. The field work would provide useful findings about the performance analyst's role that could confirm or challenge the traditional view of performance analysis in its coaching context (Franks et al., 1983). The methods chapter of such a dissertation could include sections on:

- Research design: this is an overall summary of the methods.
- Ethical issues: the student can use a subsection of the methods chapter to describe the ethical issues involved with the research project, how entry into the field was negotiated, different ethical dilemmas that occurred during the field study and how these were managed.
- Reflexivity: the student acknowledges their own background, experience, beliefs and attitude relating to performance analysis support.
- Description of the particular case: this should include information on the make-up of the squad, the gender, level and age group of the players, the coaching structure and other management and support staff within the squad.
- Detail of the research setting: this should include details of match situations, analysis locations, training venues and briefing venues.
- Data gathering: this section should describe the methods used to record field notes and discuss any problems or issues that arose during field work. Appendices may be used to show examples of field note recording forms.
- Data analysis: there are general descriptions of qualitative data analysis methods in research methods textbooks (for example Gratton and Jones, 2004: 220–21). However, students should describe what they actually did during data analysis and the unique aspects of their project.
- Data reporting: once the data have been analysed, there is still a process of transforming the emergent coding structures within the data into the forms in which they will be reported. The student should describe any such processes that the reader might not appreciate just from reading the results and discussion chapters.

There may also be research projects into theoretical aspects of performance analysis that are purely interview-based. Explanations of interacting performances theory (O'Donoghue, 2009a), scoreline effects or venue effects could come from interviews with coaches or athletes. The methods chapter of such a dissertation should include subsections on research design, ethical issues, interview guide, participants, the actual interviewing, establishing trustworthiness and data analysis. Again, although there are good sections of text books on designing and carrying out interview studies (Gratton and Jones, 2004: 144–54), the student should describe the unique aspects of their own interview study.

Where qualitative techniques are used to support quantitative techniques within a mixed methods approach, the use of qualitative methods may be described within a single section that covers participants, data gathering, data analysis, trustworthiness and validity, or these aspects of the qualitative methods used may require separate sections. Where the use of both the quantitative and qualitative methods requires different subsections, it is best to first structure the methods into different sections for research design, the quantitative methods and then the qualitative methods before breaking the quantitative and qualitative methods sections into subsections. Most dissertations of this kind would use a different heading for the qualitative methods section, such as 'interview study' or 'coach evaluation' that reflects the role of the qualitative methods within the research design. Similarly, the heading of the quantitative section would reflect the role of quantitative methods within the research design and might be entitled 'match analysis' or 'performance profiling', for example.

Quantitative results

There is usually a single results chapter within a dissertation that includes all of the findings, whether quantitative or qualitative. In the current section, the presentation of quantitative results is covered. In quantitative research, summary results should be presented allowing the different samples to be compared. A reductive approach is used to represent the samples by sample parameters, and individual performance results should not be included. The presentation of numerous similar charts (or tables) for different performances demonstrates a failure by the student to understand the nature of quantitative research, data analysis and an inability to present a concise set of results that answer the research question.

In studies where differences between samples are being tested, the student should use sample parameters to summarise the whole sample; these could be means or medians to represent the sample average with supporting measures of dispersion such as the standard deviation, range or the inter-quartile range. These descriptive statistics should be presented in text, in charts or in tables. The student should never present the exact same results in both a table and a chart because this confuses the reader who is expecting each table or figure to represent a different result. Similarly, the text of a results chapter should not merely repeat the results shown in tables and charts but should draw the readers' attention to patterns within and across different results.

In studies where relationships between variables are being tested, charts and supporting correlation and possibly regression results are presented for relationships between numerical variables. Cross-tabulated frequencies are used to show the association between categorical variables. Performance analysis investigations can also use results types that are unique for the discipline. For example, biomechanical analysis studies can use 'stick figure' representations to show joint angles and other kinematic and kinetic results at key points in the performance of a skill. Notational analysis studies can display results comparing performance in different areas of the playing surface using graphical or text form within diagrams of the playing surface. Whatever form of descriptive and inferential results are produced, each result should play a role in answering the research question. For example, if we are comparing performances between home and away matches, venue should be a variable within each result type.

Tables

Tables are useful for summarising large amounts of data. Tables should have a title and any notes required to ensure the reader can understand the content. The column and row headings should include any units of measurements that apply to the results presented in the main body of the table. The style of presentation should be consistent and should follow guidelines given in the module handbook for the dissertation module.

There are different types of table that can be used to summarise results, including univariate, bivariate and multivariate tables. Univariate tables summarise a single variable as a frequency distribution. This type of table might be useful in single-sample studies but in studies comparing samples, each result should have some role in answering the research question. Bivariate tables summarise two variables, one of which could be the grouping variable used to distinguish samples while the other would be some dependent variable of interest. Bivariate tables are useful for showing cross-

Table 10.1 Number of passes made by netball players per match (values in parentheses are percentages for the type of performance) (fictitious data)

Position	Type of Performance				
	Top v Top (n=4)	Top v Bottom (n=4)	Bottom v Top (n=4)	Bottom v Bottom (n=4)	All (n=16)
GK	39.5 (8.5%)	32.0 (6.6%)	35.8 (8.5%)	27.5 (5.8%)	33.7 (7.3%)
GD	49.3 (10.5%)	43.8 (9.0%)	55.3 (13.1%)	51.3 (10.9%)	49.9 (10.8%)
WD	66.3 (14.2%)	58.0 (12.0%)	53.5 (12.7%)	48.3 (10.3%)	56.5 (12.3%)
C	115.3 (24.7%)	129.3 (26.7%)	114.8 (27.2%)	136.3 (29.0%)	123.9 (26.9%)
WA	96.8 (20.7%)	103.3 (21.3%)	86.8 (20.6%)	99.0 (21.1%)	96.4 (20.9%)
GA	73.8 (15.8%)	80.8 (16.7%)	55.5 (13.2%)	77.8 (16.5%)	71.9 (15.6%)
GS	26.5 (5.7%)	37.3 (7.7%)	20.5 (4.9%)	30.3 (6.4%)	28.6 (6.2%)
Total	467.3 (100.0%)	484.3 (100.0%)	422.0 (100.0%)	470.3 (100.0%)	460.9 (100.0%)

Key:
Top   Team in the top half of the league
Bottom   Team in the bottom half of the league

tabulated frequencies of categorical variables and when the unit of analysis is a match event rather than whole performances. Table 10.1 is a fictitious example of a bivariate table showing the players who made passes in four different types of netball performance. Note that in this example, the percentages are of the column totals so that the percentage distribution can be seen for each type of performance.

Most tables used in performance analysis are multivariate tables where the grouping variable (or variable distinguishing related samples) is represented by the columns, and the rows are used to represent a set of hypothesised dependent variables. Table 10.2 is an example of such a table that shows fictitious data for the percentage of tennis points of different types. Note that this table augments the inferential statistics results to the descriptive statistics so space is not required in the text of the results chapter to cover these.

Thomas *et al.* (2005: 379–86) gave general examples of good and poor practice when presenting tables and charts. The main areas of poor practice to be avoided when producing tables are:

• Using a table for results that could effectively be presented in a single sentence of text within the results chapter.
• Using a table to present post hoc tests, whether parametric or nonparametric, that could easily be augmented with a table or chart showing the corresponding descriptive results.
• Using a table where a chart would have greater visual impact.
• Using inconsistent numbers of decimal places to present values in the same row of a table where the row represents the same variable. If a column represents a single variable then a consistent number of decimal places should also be used within the column.

Charts

Well presented charts have visual impact when displaying results to an audience. Kumar (2005: 252–6) showed good examples of bar charts where the values represented by bars are also shown. However, precise descriptive statistics might not be displayed if error bars are shown as well. Students wishing to compare their results with the results of previous research are often frustrated that previously published research uses charts that do not show the precise value of descriptive statistics of interest. Charts should have titles, legends, any necessary axes titles and values to aid their clarity. There are various types of chart that can be used and each is suited to different situations.

The histogram is a bar graph showing the frequency distribution of some continuous variable. The histogram uses a bar for each sub-range of values and there are no gaps between the bars to reflect the fact that there are no values of the continuous variable between the values of successive bars. In

Table 10.2 Percentage of different types of points played at different tournaments (fictitious data)

Point type	Tournament				Kruskal Wallis H test
	Australian Open (n=32)	French Open (n=29)	Wimbledon (n=44)	US Open (n=39)	
Ace	5.5±3.2	4.3±2.9 ^	7.3±4.5	6.0±3.8	p = 0/016
Double Fault	4.1±2.8	3.4±2.5	3.8±2.9	3.6±1.9	p = 0.729
Serve Winner	16.4±4.2 ^	15.1±3.9 ^$	22.8±7.1	19.5±6.9	p < 0.001
Return Winner	2.1±1.7	1.6±1.5	1.6±1.6	1.8±1.5	p = 0.384
Server to net first	9.0±5.2 ^$	11.0±3.3	13.1±4.9	12.3±4.2	p = 0.001
Receiver to net first	4.6±2.7	6.0±2.5	5.1±2.7	6.1±2.8	p = 0.079
Baseline rally	58.3±8.6 ^$	58.6±5.2 ^$	46.3±9.4	50.7±10.7	p < 0.001

Key:
^ Bonferroni adjusted post hoc test revealed significant difference to Wimbledon (p < 0.008)
$ Bonferroni adjusted post hoc test revealed significant difference to US Open (p < 0.008).

performance analysis, a variation of the histogram can be used to represent mean frequencies or percentages for a sample rather than the frequency or percentage of cases. This also allows error bars to show the variation about the mean. Figure 10.1 is an example of such a chart, showing fictitious results for heart rate response during a field game. This chart is technically not a histogram, but there are no gaps between the bars as the percentage heart rate max is still a continuous variable. The error bars here represent standard deviations, but error bars could also be used to represent standard errors or confidence intervals.

The simple bar graph shows the value of a single variable for each sample in the study. An example is shown in Figure 10.2, which displays the percentage of time spent performing high intensity activity during the four quarters of a sample of fictitious netball matches. When one considers that a single row of a table could display these means and standard deviations with further rows being used for other variables, it is not recommended that students use several simple bar charts instead of a single table. Where the simple bar chart is useful is when a critically important dependent variable of the study should be displayed in isolation. The results of post hoc tests are also shown on the chart, meaning that the text only has to report the main repeated measures ANOVA (or Friedman test) result. The charts shown in Figures 10.1 and 10.2 have only one type of bar each (5 per cent sub-range of percentage heart rate max and quarter of a netball match respectively). Therefore, a legend should not be used as it is already obvious what the bars represent.

The clustered bar chart is used to represent values for more than one independent factor. For example, Figure 10.3 shows a clustered bar chart representing the mean number of shots played per second in fictitious men's

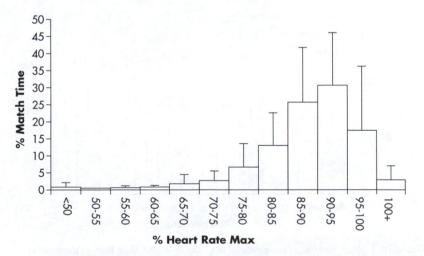

Figure 10.1 Distribution of %HR max during match time in competitive netball (fictitious data)

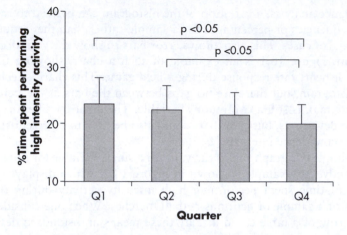

Figure 10.2 Proportion of high intensity activity performed in different quarters of a netball match (fictitious data)

and women's singles matches at four different Grand Slam tennis tournaments. The clustered bar chart, like the simple bar chart, can be used with or without error bars. It is essential to use a legend in a clustered bar graph to distinguish the meaning of the different types of bar. This bar graph could also have inferential statistics added, although the choice as to whether to include these in the chart or the accompanying text should be delayed until the inferential tests are done. If there is no significant gender, tournament or interaction effects then this can be stated within the accompanying text of the results chapter. If there are multiple significant differences, these may clutter up the chart if included. The student should sketch out the chart

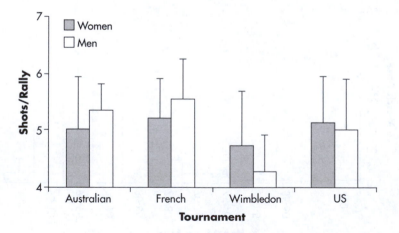

Figure 10.3 Mean number of shots played per rally in Grand Slam tennis tournaments (fictitious data)

and look at the different ways of presenting post hoc test results before committing to adding them to the chart or keeping them separate.

The stacked bar chart places bars on top of each other when the total of the values represented by the bars has meaning in addition to the individual bars. For example, Figure 10.4 shows the distance covered during a netball match by fictitious players of different positions using different locomotive movements. The total of the bars represents the total distance covered during a match using any locomotive movement.

A line graph is used to represent the influence of one continuous variable on another. It is important to recognise that unlike the scattergram, there is a set of values for the independent variable for which data have been collected from all units of analysis (subjects of the study). Figure 10.5 is an example of 100m split times during 1,500m heats, semi-finals and finals during a fictitious athletics competition. This type of graph can be used to show trends in performance indicators over a period of time either within or between matches. In biomechanical analysis of technique, line graphs are excellent for showing how displacement, velocity, acceleration, angle, angular velocities and angular acceleration change with time during the performance of a skill.

The scattergram is used to plot two continuous variables against each other to display the strength of their relationship. There will be a dot on the chart for each subject (match, performance, player or other unit of analysis) using the values of the pair of variables being displayed. For example, Figure 10.6 shows the relationship between the percentage of points where the first serve is in and the percentage of points won when the first serve is in. Scattergrams can also have regression lines drawn through them if the relationship between the two variables is strong enough.

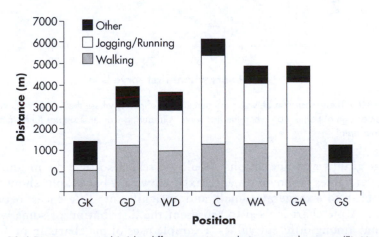

Figure 10.4 Distance covered (m) by different positions during a match quarter (Fictitious data)

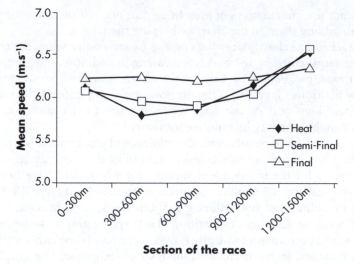

Figure 10.5 Mean running speed for 300m sections of international women's 1,500m races (Fictitious data)

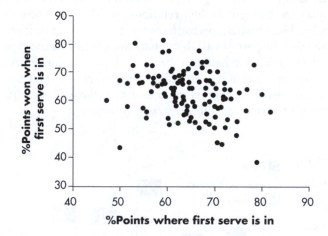

Figure 10.6 The association between the percentage of points where the first serve is in and the percentage of points won when the first serve is in during women's singles tennis matches (Fictitious data)

Bar graphs and line graphs should be scaled so that the meaningful range of values is shown on each axis. However, bar graphs should not be scaled to exaggerate small and insignificant differences between values. A pie chart is used to represent the distribution of some whole amount among different values. Example uses of pie charts in performance analysis are:

- the distribution of time among different movement classes;
- the distribution of time among different behaviours;
- the distribution of distance covered among different movement classes;
- the distribution of match events among different event types.

Figure 10.7 shows the breakdown of match time within fictitious tennis matches between rally time, inter-serve time, inter-point time within games and inter-game time when changing ends or not. Pie charts are not good at comparing samples; for example if we wished to compare the breakdown of coaching session time between three different levels of coach, a single stacked bar chart would be more concise than three separate pie charts. Where pie charts are being produced for theses or papers where colour is not permitted, labelling the values as in Figure 10.7 should be used rather than using similar shades of grey.

Some students use three dimensional representations of charts. Students should remember that the main purpose of a chart is to convey a finding of the study. Therefore, when using three dimensional versions of charts, the student should ensure that this does not compromise the reader's understanding of the values being displayed.

Playing surface diagrams

In performance analysis there are many results that use a bird's eye view of the playing area to display the location of events. Hughes and Franks (2004c) showed examples of playing area diagrams for soccer, field hockey and squash, and Hughes and Franks (2004d) showed examples of playing surface diagrams for tennis, basketball and netball. Figure 10.8 is an example

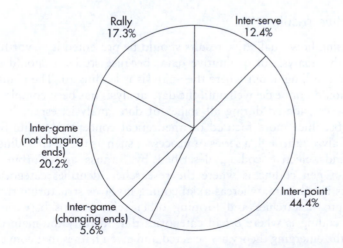

Figure 10.7 Distribution of match time during Grand Slam singles tennis (Fictitious data)

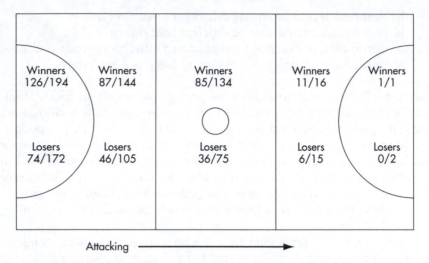

Figure 10.8 Fraction of possessions starting in different court areas that lead to goals in international netball (data from Williams and O'Donoghue, 2006)

of a playing surface diagram for fictitious netball matches, which shows the fraction of possessions starting in different areas that led to goals being scored. There are five different areas and results for winning and losing teams within matches are shown. These types of diagrams can be used for different types of playing areas in different sports to show the volume of events as well as the effectiveness of events. Some diagrams may be parts of playing surfaces, for example in a study of tennis service it may only be necessary to show the service courts.

Qualitative results

In discussing how qualitative results should be presented it is worth recapping on the analysis of qualitative data, because students should analyse their data thinking about where the analysis is leading to. The planning of results should not be delayed until the data analysis has been completed, but should be considered during all stages of data analysis, especially as the analysis becomes more focused on theoretical concepts. Qualitative data can be analysed through a series of processes such as the open coding, axial coding and selective coding, described by Strauss and Corbin (1998: 101–61). Open coding is where the researcher identifies categories and properties in the data, whereas axial coding involves structuring these categories into hierarchies and forming other relationships between them. Selective coding is where broad categories that identify the main components of the emerging theory are selected and given names that convey these components at a high level of abstraction.

There are several different approaches to writing up the findings of qualitative research (Kvale and Brinkmann, 2009: 284–90). Two approaches that are particularly relevant for qualitative research related to performance analysis are narratives and visualisation. The narrative approach involves presenting the results as a story or as a series of stories describing the findings. The results can show quotations from interview data that characterise different components of the theory. The best quotes should be used and attributed to the interviewees using a simple signature scheme that conceals their true identities (Kvale and Brinkman, 2009: 279–81). There should be a balance between the use of quotes and text in the results, with the text explaining the questions that gave rise to particular quotes shown and interpreting those quotes in terms of the theory and its components.

The student should read their results and also examine how many quotes are derived from individual participants. It is common for many of the best quotes to come from one or two participants out of a set of 10 or more. The results may need to compromise between using the best quotes and using a spread of quotes from different participants in order to demonstrate that the emerging theory is based on the full sample rather than being specific to one participant's experience or views. Excerpts of the researcher's detailed field notes can also be included in results chapters of qualitative research studies. McLaughlin and O'Donoghue (2004) used excerpts from McLaughlin's field notes to convey the evidence supporting the five different themes emerging from the data. The text of the results mediated between these real-world observations and the abstract theory being presented.

Visualisation can involve the use of tree diagrams to show structures of components of the theory, or flow diagrams to show temporal, causal or logical relations between components. The coaching model (Côté et al., 1995) is a good example of the use of a diagram to portray the components of a theory. The purpose of such diagrams is to represent the views of informants in a way that has visual impact. The presentation of results needs to engage the reader and there is often a conflict between wanting to write scientifically and wanting to write in a more popular form (Kvale and Brinkman, 2009: 266). The writing style is particularly important in areas of sports science related to performance analysis because the readers will not only be scientifically knowledgeable, but will genuinely be interested in sport and coaching. While trying to make the results interesting, the results should not include speculations that are not supported by the evidence in the data. The results should also describe variability in the theory where it has been necessary to recognise this as a result of negative cases found in the data (Strauss and Corbin, 1998: 159–61).

The discussion

The discussion chapter is usually allocated more marks than the results but takes much less time to complete. Producing the results involves the collection and analysis of data as well as writing the chapter. What is being assessed in the discussion chapter is the student's knowledge and intellectual ability that have developed over the course of their university degree programme. The student needs to be able to explain their findings in relation to theory from performance analysis, coaching and other disciplines that are relevant to the project. The discussion should not merely repeat the results in words but should explain the results and discuss the implications of the results through logical reasoning. For example, the results of a time-motion analysis study can be considered with the results from published physiology studies to make an inference about energy system utilisation and fatigue in the particular sport. Similarly, a study of coaching style should consider the style observed for the participant coaches together with teaching and coaching literature about strengths and weaknesses of different approaches when used with different genders and age groups. A study of tactical differences between different levels of teams in a field game should draw upon coaching literature for that team game to discuss the implications of the tactics used and offer explanations as to why the differences observed are associated with quality of performance.

The discussion should not follow a result-by-result format as the explanations and implications of some areas may be similar. Therefore, the discussion chapter should be organised into broad areas of the research problem that can be discussed in a more holistic way using the evidence from the study and any relevant literature. The discussion should cover where the results support or challenge the outcomes of previous research and offer potential reasons for any disagreements that are found. Before completing the discussion chapter, the student should read and re-read the chapter to ensure it is well written and structured. A simple check may reveal five pages in a row that do not refer to any academic literature. The student should revise their discussion in such sections to include literature that supports the explanations and arguments being made.

The conclusions

The conclusions chapter should include the summary findings of the study, practical recommendations and areas for future work that are possible to develop the research area further. The conclusions summarise the main findings of the study without any further discussion. These findings provide the answer to the research question posed in the introduction chapter. The student should report what they found and not what they would have wished to find. The practical recommendations must also be supported by the findings of the study and students should not make unsubstantiated claims that cannot be supported. For example, an observational study will provide useful results that coaches should be aware of, but it does not provide any evidence that specific preparation based on the findings will actually be effective.

Areas for future research should not be presented as things that were not done in the student's project, but instead they should be listed as opportunities made possible by the advances made within the student's research. The restricted scope of any research project gives ample possibilities for future study. Some common areas of future research listed in performance analysis dissertations are the application of alternative methods, investigating different levels of performance within the same sport, and performing follow-up experimental investigations.

Completing the dissertation

In addition to the main chapters there is other material that needs to be included within the dissertation. This includes the abstract, acknowledgements, contents, lists of figures and tables and any appendices. The abstract is an overall synopsis of the research, describing its purpose, general methods, main findings and implications, using between 100 and 200 words.

Acknowledgements should be made to anybody who assisted the student in completing the project, especially any participants and colleagues who assisted the student in performing an inter-operator reliability test for the system. The support of the supervisor, other academic staff and technicians is also included. Very often students also thank family and friends who have supported them during the process of doing the research project as well as during the degree of which the dissertation is a part.

Appendices do not need to include all of the raw data captured by the students as this can be maintained electronically or within a verification file. Appendices may include pre-processed results, the output from statistics packages, blank data collection forms, summary analysis sheets, screen dumps of computerised systems, ethics forms and any letters requesting permission to film or any other voluntary informed consent forms.

WRITING UP SEPARATE STUDIES OF A PHD THESIS

Performance analysis of sport is a relatively new discipline, but already PhD research has been successfully completed using performance analysis techniques. Some PhD studies have used performance analysis as one of a number of methods to investigate the chosen research topic (Bloomfield, 2005, McLaughlin, 2003) with others exclusively using performance analysis (Boddington, 2002, Choi, 2008). A PhD typically involves a series of three to five studies that address different aspects of some research topic of interest. The contents of such a PhD thesis would typically be:

Introduction
Literature review
Methodology
A chapter for each study (each of these chapters would include the purpose, methods, results and brief summary conclusions)
Overall discussion
Conclusions

The methodology at this level would be expected to discuss ontological and epistemological aspects of the research. There may be some common methods used within all of the studies in the PhD that should be described before the individual studies. A PhD in performance analysis might include studies on the validity of performance indicators, their reliability, a study comparing samples related to some independent factors of interest, and an intervention study where key performance indicators are evaluated within coaching practice. Studies identifying valid performance indicators could use a combination of quantitative techniques and focus groups. The intervention study is typically a case study applying some novel approach developed during the research and evaluating its effectiveness using a combination of self-report data, field notes and performance data.

OTHER TYPES OF RESEARCH REPORT

Writing a conference abstract

The abstract for a conference is similar to the abstract for a dissertation in that it is an overall synopsis of what was done and what was found. However, an abstract for a conference might include a summary table of results as well as give more detail to a conference's scientific committee who will not have the benefit of being able to consult the student's dissertation.

Preparing slides for a conference presentation

If the student's work is accepted for podium presentation at a conference, they need to produce a set of slides. A conference presentation may be as short as 15 minutes including questions, so the student should include key slides on the background, purpose, methods, results, discussion and conclusions. Given that most academics realise that research presentations include these parts, it is not recommended that students take up a minute of their presentation with an overview slide. Slides should not contain a large volume of unreadable text but should include key points discussed at each stage. Charts that clearly indicate what has been found should be presented in the results. Other images of equipment can be included in the methods slides.

The student should not use a script as this can interrupt their flow during the presentation and the student should be an authority on their own work, able to talk naturally about it without prompting. The student should also be able to sum up what they did, what they found and what it means in less than a minute, just in case they run out of time during the presentation. This is always a possibility, even if the student has been well prepared, as the questioning of a previous presentation may have put the conference behind schedule if the chair did not keep good control of the session.

Posters

Research can also be presented in the form of a poster, which can include more detail than would be presented on slides. The poster should include key figures, tables and photographs to convey details of the methods, results and theoretical aspects. The poster can be viewed by others at their leisure and discussed with the student informally. Some conferences give poster presenters three minutes to describe what they did and what they found and answer one or two questions.

Writing a journal paper

Student research has occasionally been of such an exceptional quality that it has been published in peer reviewed sports science journals. The process of producing a paper typically involves cutting the dissertation down into the format of a paper required by the journal it is being submitted to. Different journals have different guidelines for authors, with some using a basic introduction to the research problem and the purpose of the research, while others allow a greater coverage of background material in the introduction section. Any student seeking to have their work published in a journal should look at the types of papers that are published within the journal and understand its scope. The supervisor should be able to assist in the development of the paper, especially if the supervisor is a research-active member of staff.

A key point made earlier in this book was that the student is at university to work towards their degree qualification and should be aware that efforts towards getting their research published may be at the expense of their degree performance. Therefore, the process of publishing a paper would typically commence after the student has completed their programme of study.

REFERENCING

There are various referencing systems in use, such as the Harvard system used in this textbook. The student should consistently use the referencing system specified for dissertations within their own programme of study. Close to completion of the dissertation, the student should check that all references used in the text are in their reference list and that all references in the list are actually used somewhere within the text of the dissertation. Students should develop their reference list while doing their literature search to ease the task of finalising the list close to the dissertation submission date.

OTHER CONSIDERATIONS

Students should consult the dissertation handbook and other support material that is available to help them do their project and complete the dissertation. The supervisor should be consulted throughout the project and reporting should be regular and detailed. Where a student's work is delayed until close to the submission date for the dissertation, the supervisor may not be available to see the student as often as the student would like. Supervisors can read draft chapters but would be reluctant to give repeated feedback on a chapter to the extent that it was no longer the work of the student. Usually, supervisors do not read the discussion chapter of the dissertation until the final thesis is submitted.

SUMMARY

There are various types of research reports that are written for different audiences. The research proposal is important as it is a means by which a student can identify a research project of interest, have a supervisor allocated and be granted ethical approval to commence their research project. This chapter has covered issues in the write-up of the different chapters to be produced within the thesis. This author has experienced many occasions where students have produced theses that have not done justice to the research they have done. Students doing performance analysis dissertations not only need to do good technical work but also need to do good academic work and present their work well to be awarded high marks.

REFERENCES

Altman, D.G. (1991) *Practical Statistics for Medical Research*, London: Chapman & Hall.

Ariel, G. (2007) 'Sports technologies from Mexico City Olympics to the future Olympics in Beijing 2008', paper presented at the 6th International Symposium of the International Association of Computer Science in Sport, Calgary, June.

Armitage, P.J. (2006) An Analysis of the Knock Out Stages of the 2003 Rugby Union World Cup. Unpublished BSc Hons thesis, University of Wales Institute Cardiff.

Atkinson, G. (2002) 'What is this thing called measurement error?', paper presented at the 12th Commonwealth International Sport Conference, Manchester, July.

Atkinson, G. and Nevill, A.M. (1998) 'Statistical methods for addressing measurement error (reliability) in variables relevant to sports medicine', *Sports Medicine*, 26: 217–38.

Australian Open (2009) 'Official website of the Australian Open', Available <http://www.ausopen.org> (accessed 27 July 2009).

Baker, J., Ramsbottom, R. and Hazeldine, R. (1993) 'Maximal shuttle performance over 40m as a measure of anaerobic performance', *British Journal of Sports Medicine*, 27: 228–32.

Balague, N., McGarry, T., Lames, M., Palut, Y. and Walter, F. (2005) 'Modelling the interaction in game sports by relative phase', paper presented at the 5th International Symposium of Computer Science in Sport, Hvar, Croatia, May.

Bandura, A. (1977) 'Self-efficacy: toward a unifying theory of behavioural change', *Psychology Review*, 84: 191–215

Bangsbo, J. (1993) *The Physiology of Soccer – With special reference to intense intermittent exercise*, Copenhagen: HO & Strom.

Bangsbo, J. (1997) The physiology of intermittent activity in football. In T. Reilly, J. Bangsbo and M. Hughes (eds), *Science and Football 3* (pp. 43–53). London: E & FN Spon.

Bangsbo, J., Nørregaard, L. and Thorsøe, F. (1991) 'Activity profile of professional soccer', *Canadian Journal of Sports Sciences*, 16: 110–16.

Barbacci, S. (2002) 'Labanotation: a universal movement notation language', *Journal of Science Communication*, 1: 1–11.

Bartlett, R.M. (1999) *Sports Biomechanics: Reducing injury and improving performance*, London: Routledge.

Bartlett, R. (2001) 'Performance analysis: can bringing together biomechanics and notational analysis benefit coaches?', *International Journal of Performance Analysis in Sport*, 1: 122–26.

Beck, C. and O'Donoghue, P.G. (2004) Time-motion analysis of intervarsity rugby league competition. In O'Donoghue, P.G. and Hughes, M. (eds), *Performance Analysis of Sport 6* (pp. 150–55). Cardiff: CPA Press, UWIC.

Bland, J.M. and Altman, D.G. (1986) 'Statistical methods for assessing the agreement between two methods of clinical measurement', *Lancet*, I: 307–10.

Blaze, A., Atkinson, G., Harwood, C. and Cale, A. (2004) Prevalence and perceptions of performance analysis in the English Football Association Premier League. In O'Donoghue, P.G. and Hughes, M. (eds) *Performance Analysis of Sport 6* (pp. 79–83). Cardiff: CPA Press, UWIC.

Bloom, G., Crumpton, R. and Anderson, J. (1999) 'A systematic observation study of the teaching behaviours of an expert basketball coach', *The Sports Psychologist*, 13: 157–70.

Bloomfield, J. (2005) FA Premier League soccer: physical characteristics, physical demands of match-play and effective physical conditioning. Unpublished PhD thesis, University of Hull.

Bloomfield, J., Polman, R. and O'Donoghue, P.G. (2004) 'The Bloomfield Movement Classification: movement analysis of individual players in dynamic movement sports', *International Journal of Performance Analysis of Sport*, 4(2): 20–31.

Bloomfield, J., Polman, R. and O'Donoghue, P.G. (2007a) 'Reliability of the Bloomfield Movement Classification', *International Journal of Performance Analysis of Sport*, 7(1): 20–27.

Bloomfield, J., Polman, R., O'Donoghue, P. and McNaughton, L. (2007b) 'Effective speed and agility conditioning methodology for random intermittent dynamic type sports', *Journal of Strength and Conditioning Research*, 21: 1093–100.

Blucher, S. and O'Donoghue, P.G. (2007) 'A technique to interpret performance indicators norms' paper presented at the 4th All Wales Sport, Exercise Science and Medicine Conference, Glamorgan, July.

Boddington, M.K. (2002) The efficacy of visual feedback to enhance sporting performance, with specific reference to field hockey. Unpublished PhD thesis, University of Cape Town.

Boffin, D. (2004) An analysis of the nature of aggressive play on clay in elite male tennis players. Unpublished BSc Hons Thesis, University of Wales Institute Cardiff.

Borg, G.V. (1982) 'Psychological basis of perceived exertion', *Medicine and Science in Sports and Exercise*, 14: 377–81.

Borrie, A. (1997) 'When is research not research – a response', *BASES Newsletter*, 7(7): 5–6.

Brown, E. (2005) 'Running strategy of female middle distance runners attempting the 800m and 1500m "Double" at a major championship: a performance analysis and qualitative investigation', *International Journal of Performance Analysis of Sport*, 5(3): 73–88.

Brown, D. and Hughes, M. (1995) The effectiveness of quantitative and qualitative feedback on performance in squash. In T. Reilly, M. Hughes and A. Lees (eds), *Science and Racket Sport* (pp. 232–7). London: E and FN Spon.

Brown, E. and O'Donoghue, P.G. (2006) Analysis of performance in running events. In H. Dancs, M. Hughes and P.G. O'Donoghue (eds), *Performance Analysis of Sport 7* (pp. 337-348). Cardiff: CPA Press, UWIC.

Brown, E. and O'Donoghue, P.G. (2007) 'Relating reliability to analytical goals in performance analysis', *International Journal of Performance Analysis of Sport*, 7(1): 28–34.

Brown, E. and O'Donoghue, P.G. (2008a) 'A split screen system to analyse coach behaviour: a case report of coaching practice', *International Journal of Computer Science in Sport*, 7(1): 4–17.

Brown, E. and O'Donoghue, P.G. (2008b) 'Gender and Surface effect on elite tennis strategy', *Coaching and Sports Science Review*, 46: 9–11, Available <http://www.itftennis.com/shared/medialibrary/pdf/original/IO_38643_original.PDF> (accessed 25 June 2009).

Brown, E. and O'Donoghue, P.G. (2008c) 'Gender and surface effect on rally duration in singles matches at Grand Slam tennis tournaments in 1997–9 and 2007', paper presented at the 8th World Congress of Performance Analysis of Sport, Magdeburg, September.

Burrell, G. and Morgan, G. (1979) *Sociological Paradigms and Organizational Analysis*, London: Heinemann Educational.

Carling, C., Williams, A.M. and Reilly, T. (2005) *Handbook of Soccer Match Analysis: A systematic approach to improving performance*, London: Routledge.

Carling, C., Reilly, T. and Williams, A.M. (2009) *Performance Assessment for Field Sports*, London: Routledge.

Carling, C., Bloomfield, J., Nelson, L. and Reilly, T. (2008) 'The role of motion analysis in elite soccer: contemporary performance measurement techniques and work rate data', *Sports Medicine*, 38: 839–62.

Choi, H. (2008) Definitions of performance indicators within real-time and lapse time analysis systems in performance analysis of sport. Unpublished PhD thesis, University of Wales Institute Cardiff.

Choi, H., O'Donoghue, P.G. and Hughes, M. (2006a) A study of team performance indicators by separated time scale using a real-time analysis technique within English national basketball league. In H. Dancs, M. Hughes and P.G. O'Donoghue (eds), *Performance Analysis of Sport 7* (pp. 138–41). Cardiff: CPA Press, UWIC.

Choi, H., Reed, D., O'Donoghue, P.G. and Hughes, M. (2006b) The valid numbers of performance indicators for real-time analysis using prediction models within men singles in 2005 Wimbledon Tennis Championship. In H. Dancs, M. Hughes and P.G. O'Donoghue (eds), *Performance Analysis of Sport 7* (pp. 220–26), Cardiff: CPA Press, UWIC.

Choi, H., O'Donoghue, P.G. and Hughes, M.D. (2007) 'An investigation of inter-operator reliability tests for real-time analysis system', *International Journal of Performance Analysis of Sport*, 7(1): 49–61.

Choi, H., O'Donoghue, P.G. and Hughes, M.D. (2008) A comparison of whole match and individual set data in order to identify valid performance indicators for real-time feedback in men's singles tennis matches. In A. Lees, D. Cabello and G. Torres (eds), *Science and Racket Sports 4* (pp. 227–31). London: Routledge.

Coalter, A., Ingram, B., McCrorry, P. MBE, O'Donoghue, P.G. and Scott, M. (1998) 'A comparison of alternative operation schemes for the computerised scoring system for amateur boxing', *Journal of Sports Sciences*, 16, 16–17.

Cohen, J. (1960) 'A coefficient of agreement for nominal scales', *Educational and Psychological Measurement*, 20: 37–46.

Cohen, J. (1968) 'Weighted kappa: nominal scale agreement with provision for scaled agreement or partial credit', *Psychological Bulletin*, 70: 213–20.

Cohen, L., Manion, L. and Morrison, K. (2007) *Research Methods in Education*, 6th edn, London: Routledge.

Côté, J., Salmela, J.H., Baria, A. and Russell, S. (1993) 'Organising and interpreting unstructured qualitative data', *The Sports Psychologist*, 7: 127–37.

Côté, J., Trudel, P., Baria, A. and Russell, S.J. (1995) 'The coaching model: a grounded assessment of expert gymnastic coach's knowledge', *Journal of Sport and Exercise Psychology*, 17: 1–17.

Croucher, J.S. (1982) 'The effect of the tennis tie-breaker', *Research Quarterly for Exercise and Sport*, 53: 336–9.

Csataljay, G., O'Donoghue, P.G., Hughes, M. and Dancs, H. (2008) 'Valid performance indicators in basketball', paper presented at the World Congress of Performance Analysis of Sport 8, Magdeburg, September.

Cushion, C.J. and Jones, R.L. (2001) 'A systematic observation of professional top-level youth soccer coaches', *Journal of Sport Behaviour*, 24: 354–65.

Dadds, M.R., Barrett, P.M., Rapie, R.M. and Ryan, S. (1996) 'Family process and child anxiety and aggression: an observational analysis', *Journal of Child Psychology*, 24: 187–203.

Darnell, E., Robinson, G. and O'Donoghue, P.G. (2008) 'Injury risk during netball competition: an observational investigation', paper presented at the World Congress of Performance Analysis of Sport 8, Magdeburg, September.

Davies, M.J., Bloom, G.A. and Salmela, J.H. (2005) 'Job satisfaction of accomplished male university basketball coaches: the Canadian context', *International Journal of Sport Psychology*, 36: 173–92.

Devlin, G., Brennan, D. and O'Donoghue, P.G. (2004) Time-motion analysis of work-rate during home and away matches in collegiate basketball. In P.G. O'Donoghue and M. Hughes (eds), *Performance Analysis of Sport 6* (pp. 174–8). Cardiff: CPA Press, UWIC.

Dingwell, J.B., Cusumano, J.P., Sternad, D. and Cavanagh, P.R. (1999) 'Stride to stride variability in human walking is not noise', paper presented at the 23rd Annual Meeting of the American Society of Biomechanics, Pittsburgh, October.

Di Salvo, V., Gregson, W., Atkinson, G., Tordoff, P. and Drust, B. (2009) 'Analysis of high intensity activity in Premier League soccer', *International Journal of Sports Medicine*, 30: 205–12.

Donnelly, C. and O'Donoghue, P.G. (2008) 'Behaviour of netball coaches of different levels' paper presented at the World Congress of Performance Analysis of Sport 8, Magdeburg, September.

D'Ottavio, S. and Castagna, C. (2001) 'Physiological load imposed on elite soccer referees during actual match play', *Journal of Sports Medicine and Physical Fitness*, 41: 27–32.

Dreckmann, C., Görsdorf, K. and Lames, M. (2009) 'That's the way, aha aha, I like it – qualitative methods in game analysis', paper presented at the 3rd International Workshop of the International Society of Performance Analysis of Sport, Lincoln, April.

Durand-Bush, N. (1996) Training blood sweat and tears. In J.H. Salmela (ed), *Great Job Coach! Getting the edge from proven winners* (pp.102–37). Ottawa, ON: Potentium.

Elliott, B., Burnett, A., Stockhill, N. and Bartlett, R.M. (1996) 'The fast bowler in cricket: a sports medicine perspective', *Sport, Exercise and Injury*, 1: 201–6.

Ericson, C.A. (2005) *Hazard Analysis Techniques for Systems Safety*, New York: Wiley.

Flick, U. (2007) *Designing Qualitative Research*, London, Sage.

Franks, I.M. (1993) 'The effects of experience on the detection and location of performance differences in a gymnastic technique', *Research Quarterly for Exercise and Sport*, 64: 227–31.

Franks, I.M. (1997) Use of feedback by coaches and players. In T. Reilly, J. Bangsbo and M. Hughes (eds), *Science and Football 3*, (pp. 267–78), London: E and FN Spon.

Franks, I.M. and Miller, G. (1991) 'Training coaches to observe and remember', *Journal of Sports Sciences*, 9: 285–97.

Franks, I.M., Goodman, D. and Miller, G. (1983) 'Analysis of performance: qualitative and quantitative', *Sports*, March.

Franks, I.M., Hodges, N. and More, K. (2001) 'Analysis of coach behaviour', *International Journal of Performance Analysis of Sport*, 1: 27–36.

French Open (2009) 'Official website of the French Open', Available <http://www.fropen.org> (accessed 27 July 2009).

Fuller, N. and Alderson, G.J.K. (1990) The development of match analysis in game sports. In *Match Analysis in Sport: A state of the art review*, Leeds: National Coaching Foundation.

Gale, D. (1971) 'Optimal strategy for serving in tennis', *Mathematics Magazine*, 5: 197–9.

Gamble, D., Young, E. and O'Donoghue, P.G. (2007) 'Activity profile and heart rate response of referees in Gaelic football', paper presented at the World Congress of Science and Football 6, Antalya, January.

Gasston, V. (2004) Performance analysis during an elite netball tournament: experiences and recommendations. In P.G. O'Donoghue and M. Hughes (eds) *Performance Analysis of Sport 6* (pp. 8–14). Cardiff: CPA Press, UWIC.

Gasston, V. and Simpson, C. (2004) 'A netball specific fitness test', *International Journal of Performance Analysis of Sport*, 4(2): 82–96.

Gerisch, G. and Reichelt, M. (1993) Computer and video aided analysis of football games. In T. Reilly, J. Clarys and A. Stibbe (eds), *Science and Football II* (pp. 167–74). London: E & FN Spon.

Gibbs, G. (2007) *Analysing Qualitative Data*, London: Sage.

Gilbert, W. and Jackson, C., (2006) *In Search of an Effective Coaching Style. Coaching the US Olympic team*, New York: Routledge.

Gilbert, W. and Trudel, P. (2004) 'Analysis of coaching science research published from 1970-2001', *Research Quarterly for Exercise and Sport*, 75: 388–99.

Goldberger, M. (1992) 'The spectrum of teaching styles: A perspective for research on teaching physical education', *Journal of Physical Education, Recreation and Dance*, 63(1): 42–6.

Goudas, M., Biddle, S., Fox, K. and Underwood, M. (1995) 'It ain't what you do, it's the way that you do it! Teaching styles affects children's motivation in track and field', *The Sports Psychologist*, 9: 254–64.

Gratton, C. and Jones, I. (2004) *Research Methods for Sports Studies*, London: Routledge.

Greene, D. (2008) Performance analysis of elite 400m hurdles races. Unpublished BA Hons thesis, University of Wales Institute Cardiff.

Greene, D., Leyshon, W. and O'Donoghue, P.G. (2008) 'Elite male 400m hurdle tactics are influenced by race leader', paper presented at the World Congress of Performance Analysis of Sport 8, Magdeburg, September.

Griffey, D.C. (1983) 'Aptitude x treatment interactions associated with student decision-making', *Journal of Teaching in Physical Education*, 3(2): 15–32.

Hale, S. (2004) Work-rate of Welsh national league players in training matches and competitive matches. In P.G. O'Donoghue and M. Hughes (eds), *Performance Analysis of Sport 6* (pp. 35–44), Cardiff: CPA Press, UWIC.

Harries-Jenkins, E. and Hughes, M. (1995) A computerised analysis of female coaching behaviour with male and female athletes. In T. Reilly, M. Hughes and A. Lees (eds), *Science and Racket Sports* (pp. 238–43), London: E & FN Spon.

Hartshorn, A. (2009) 'Performance Profiling of Elite Modern Association Football Referees', paper presented at the 3rd International Workshop of the International Society of Performance Analysis of Sport, Lincoln, England, April.

Hay, J.G. and Reid, J.G. (1988) *Anatomy, Mechanics and Human Motion*, Englewood Cliffs, NJ: Prentice Hall.

Hawkins, R.D. and Fuller, C.W. (1998) 'An examination of the frequency and severity of injuries and incidents at three levels of professional football', *British Journal of Sports Medicine*, 32: 326–32.

Hayes, M. (1997a) 'When is research not research? When it's notational analysis', *BASES Newsletter*, 7(7): 4–5.

Hayes, M. (1997b) 'Notational analysis – the right of reply', *BASES Newsletter*, 7(8): 4–5.

Heiderscheit, B.C., Hamill, J. and van Emmerik, R.E.A. (1998) 'The importance of intersegmental coordination variability during running', paper presented at the North American Congress on Biomechanics, University of Waterloo, Canada, August.

Helsen, W.F., Hodges, N.J., Van Winckel, J. and Starkes, J.L. (2000) 'The roles of talent, physical precocity and practice in the development of soccer expertise', *Journal of Sports Sciences*, 18: 727–36.

Herzog, W. (2007) Mathematically indeterminate systems. In B.M. Nigg and W. Herzog (eds), *Biomechanics of the Musculo-skeletal System*, 3rd edn (pp. 609–21), Chichester: Wiley.

Holmes, K. (2005) *Black, White and Gold – My autobiography*, London: Virgin Books Ltd.

Holt, N.L. and Mitchell, T. (2006) 'Talent development in English professional soccer', *International Journal of Sport Psychology*, 37: 77–98.

Hopkins, W.G. (2000) 'Measurement of reliability in sports medicine and science', *Sports Medicine*, 30: 1–15.

Horton, S., Baker, J. and Deakin, J. (2005) 'Experts in action: a systematic observation of 5 national team coaches', *International Journal of Sports Psychology*, 36: 299–319.

Howe, R. (2004) 'A critique of experimentialism', *Qualitative Inquiry*, 10: 42–61.

Huey, A., Morrow, P. and O'Donoghue, P.G. (2001) From time-motion analysis to specific intermittent high intensity training. In M. Hughes and I.M. Franks (eds),

Performance Analysis, Sports Science and Computers (pp. 29–34). Cardiff: CPA Press, UWIC.

Hughes M. (1986) A review of patterns of play in squash. In J. Watkins, T. Reilly and L. Burwitz (eds), *Sports Science* (pp. 363–8). London: E & FN Spon.

Hughes, M. (1993) Notational analysis of football. In T. Reilly, J. Clarys and A. Stibbe (eds) *Science and Football 2* (pp. 151–9). London: E and FN Spon.

Hughes, M. (1998) The application of notational analysis to racket sports. In A. Lees, I. Maynard, M. Hughes and T. Reilly (eds), *Science and Racket Sports 2* (pp. 211–20), London: E and FN Spon.

Hughes, M.G. (2008) Physiology for coaches. In R.L. Jones, M. Hughes and K. Kingston (eds), *An Introduction to Sports Coaching: From science and theory to practice* (pp. 126–37), London: Routledge.

Hughes, M. and Bartlett, R. (2002) 'The use of performance indicators in performance analysis', *Journal of Sports Sciences*, 20: 739–54.

Hughes, M. and Bartlett, R. (2008) What is performance analysis? In M. Hughes and I.M. Franks (eds), *Essentials of Performance Analysis: An introduction* (pp. 8–20), London: Routledge.

Hughes, M. and Blunt, R. (2001) Work-rate of rugby union referees. In M. Hughes and F. Tavares (eds), *Notational Analysis of Sport 4* (pp. 184–90). Porto: University of Porto Press.

Hughes, M. and Clarke, S. (1995) Surface effect on elite tennis strategy. In T. Reilly, M. Hughes and A. Lees (eds), *Science and Racket Sports* (pp. 272–7), London: E and FN Spon.

Hughes, M. and Franks, I.M. (1997) *Notational Analysis of Sport*, London: E and FN Spon.

Hughes, M. and Franks, I.M. (1995) 'Computerised notational analysis of football', paper presented at the World Congress of Science and Football III, Cardiff, May.

Hughes, M. and Franks, I.M. (2004a) (eds) *Notational Analysis of Sport, 2nd edn: Systems for better coaching and performance in sport*. London: Routledge.

Hughes, M. and Franks, I.M. (2004b) Sports analysis. In M. Hughes and I.M. Franks (eds), *Notational Analysis of Sport, 2nd edn: Systems for better coaching and performance in sport* (pp. 107–17). London: Routledge.

Hughes, M. and Franks, I.M. (2004c) How to develop a notation system. In M. Hughes and I.M. Franks (eds), *Notational Analysis of Sport, 2nd edn: Systems for better coaching and performance in sport* (pp. 118–40), London: Routledge.

Hughes, M. and Franks, I.M. (2004d) Examples of notation systems. In M. Hughes and I.M. Franks (eds), *Notational Analysis of Sport, 2nd edn: Systems for better coaching and performance in sport* (pp. 141–87), London: Routledge.

Hughes, M. and Franks, I.M. (2005) 'Analysis of passing sequences, shots and goals in soccer', *Journal of Sports Sciences*, 23: 509–14.

Hughes, M. and Probert, G. (2006) A technical analysis of elite male soccer players by position and success. In H. Dancs, M. Hughes and P.G. O'Donoghue (eds), *Performance Analysis of Sport 7* (pp. 89–104). Cardiff: CPA Press, UWIC.

Hughes, M. and Franks, I.M. (2008) *Essentials of Performance Analysis: an introduction*, London: Routledge.

Hughes, M., Franks, I.M. and Nagelkerke, P. (1989) 'A video system for the quantitative motion analysis of athletes in competitive sport', *Journal of Human Movement Studies*, 17: 212–27.

Hughes, M., David, R. and Dawkin, N. (2001a) Critical incidents leading to shooting in soccer. In M. Hughes (ed), *Notational Analysis of Sport 3* (pp. 10–21),
Cardiff: CPA Press, UWIC.

Hughes, M., Evans, S. and Wells, J. (2001b) 'Establishing normative profiles in performance analysis', *International Journal of Performance Analysis of Sport*, 1:
4–27.

Hughes, M., Cooper, S.M. and Nevill, A. (2004) Analysis of notation data: reliabil-
ity. In M. Hughes and I.M. Franks (eds), *Notational Analysis of Sport, 2nd edn:
Systems for better coaching and performance in sport* (pp. 189–204). London:
Routledge.

James, N. (2008) Performance analysis in the media. In M. Hughes and I.M. Franks
(eds), *The Essentials of Performance Analysis: An introduction* (pp. 243–63).
London: Routledge.

James, N., Mellalieu, S.D. and Jones, N.M.P. (2005) 'The development of position-
specific performance indicators in professional rugby union', *Journal of Sports
Sciences*, 23: 63–72.

Jenkins, R.E. (2006) Does the use of motivational videos enhance individual or team
performance in league netball? Unpublished BSc Hons thesis, University of Wales
Institute Cardiff.

Jenkins, R.E., Morgan, L. and O'Donoghue, P.G. (2007) 'A case study into the effectiveness of computerized match analysis and motivational videos within the
coaching of a league netball team', *International Journal of Performance Analysis
of Sport*, 7(2): 59–80.

Jones, M.V., Paull, G.C. and Erskine, J. (2002) 'The impact of a teams' aggressive
reputation on the decisions of association football referees', *Journal of Sports
Sciences*, 20: 991–1000.

Kahan, D. (1999) 'Coaching behaviour: a review of the systematic observation research
literature', *Applied Research in Coaching and Athletics Annual*, 14: 17–58.

King, S. and O'Donoghue, P. (2003) 'Specific high intensity training based on activity profile', *International Journal of Performance Analysis of Sport*, 3(2):
130–44.

Kingston, H. (2009) 'The effects of feed forward on the proficiency of penalty kicks
in football', paper presented at the 3rd International Workshop of the International
Society of Performance Analysis of Sport, Lincoln, April.

Kirk-Smith, M. (1998) 'Psychological issues in questionnaire based research', *Journal
of the Market Research Society*, 40: 223–26.

Knox, I. and O'Donoghue, P.G. (2001) 'The effect of positional role on work-rate in
inter-provincial and schoolboy rugby union', paper presented at the 3rd Exercise
and Sports Science Association of Ireland Annual Conference, Carlow, April.

Kumar, R. (2005) *Research Methodology: A step by step guide for beginners*, 2nd
edn, London: Sage.

Kvale, S. and Brinkmann, S. (2009) *Interviews: Learning the craft of qualitative
research interviewing*, 2nd edn, Thousand Oaks, CA: Sage.

Lacy, A.C. and Darst, P.W. (1984) 'Evolution of a systematic observation system:
The ASU coaching observation instrument', *Journal of Teaching in Physical
Education*, 3: 59–66.

Lacy, A.C. and Darst, P.W. (1985) 'Systematic observation of behaviour of winning
high school head football coaches', *Journal of Teaching in Physical Education*, 4:
256–70.

Lacy, A.C. and Darst, P.W. (1989) The Arizona State University Observation Instrument (ASUOI). In P.W. Darst, D.B. Zakrajsek and V.H. Mancini (eds), *Analysing Physical Education and Sport Instruction*, 2nd edn (pp. 369–77). Champaign, IL: Human Kinetics.

Lafont, D. (2007) 'Towards a new hitting model in tennis', *International Journal of Performance Analysis of Sport*, 7(3): 106–16.

Lafont, D. (2008) 'Gaze control during the hitting phase in tennis: a preliminary study', *International Journal of Performance Analysis of Sport*, 8(1): 85–100.

Laird, P. and Waters, L. (2008) 'Eye-witness recollection of sports coaches', *International Journal of Performance Analysis of Sport*, 8(1): 76–84.

Lames, M., Cordes, O. and Walter, F. (2009) 'Relative phase and oscillations in football', paper presented at the 3rd International Workshop of the International Society of Performance Analysis of Sport, Lincoln, April.

Leger, L.A. and Rouillard, M. (1983) 'Speed reliability of cassettes and tape players', *Canadian Journal of Applied Sport Sciences*, 8(1): 47–8.

Lees, A. (2008) Qualitative biomechanical analysis of technique. In M. Hughes and I.M. Franks (eds), *The Essentials of Performance Analysis: An introduction* (pp. 162–79). London: Routledge.

Lewis, R. (2004) *Kelly Holmes – My Olympic ten days*, London: Virgin Books Ltd.

Lincoln, Y.S. and Guba, E.G. (1985) *Naturalistic Enquiry*, Newbury Park, CA: Sage.

Loughran, B.J. and O'Donoghue, P.G. (1999) 'Time-motion analysis of work-rate in club netball', *Journal of Human Movement Studies*, 36: 37–50.

Loze, G.M., Collins, D. and Holmes, P.S. (2001) 'Pre-shot EEG alpha-power reactivity during expert air-pistol shooting: a comparison of best and worst shots', *Journal of Sports Sciences*, 19: 727–33.

Lyons, K. (1988) *Using Video in Sport*, Huddersfield, UK: Springfield Books.

Lyons, K. (1998) 'Origins of notational analysis: Australian roots in sports coach', paper presented at the World Congress of Notational Analysis of Sport 4, Porto, September.

McAuley, E. (1992) Self-referent thought in sport and physical activity. In T.S. Horn (ed), *Advances in Sport Psychology* (pp. 101–18). Champaign, IL: Human Kinetics.

McBride, R.E. (1992) 'Critical thinking – an overview with implications for physical education', *Journal of Teaching in Physical Education*, 11(2): 112–25.

McCorry, M., Saunders, E.D., O'Donoghue, P.G. and Murphy, M.H. (1996) A match analysis of the knockout stages of the 3rd Rugby Union World Cup. In M. Hughes (ed) *Notational Analysis of Sport 3* (pp. 230–39). Cardiff: CPA Press, UWIC.

McErlean, C.A., Cassidy, J. and O'Donoghue, P.G. (2000) 'Time-motion analysis of gender and positional effect on work-rate in Gaelic football', *Journal of Human Movement Studies*, 38: 269–86.

McGarry, T. (2006) 'Identifying patterns in squash contests using dynamical analysis and human perception', *International Journal of Performance Analysis of Sport*, 6(2): 134–47.

McGarry, T. and Franks, I.M. (1994) 'Stochastic approaches to predicting competition squash match play', *Journal of Sports Sciences*, 12: 573–884.

McGarry, T., Khan, M.A. and Franks, I.M. (1999) 'On the presence and absence of behavioural traits in sport: an example from championship squash match-play', *Journal of Sports Science*, 17: 297–311.

McGarry, T., Anderson, D.I., Wallace, S.A., Hughes, M.D. and Franks, I.M. (2002) 'Sport competition as a dynamical self-organizing system', *Journal of Sport Sciences*, 20: 771–81.

McGinnis, P.M. (1999) *Biomechanics of Sport and Exercise*, Champaign, IL: Human Kinetics.

McLaughlin, E. (2003) Children's recreational activity and health: a time-motion study. Unpublished PhD thesis, University of Ulster.

McLaughlin, E. and O'Donoghue, P.G. (2002) Activity profile of primary school children in the playground, *Journal of Human Movement Studies*, 42, 91–108.

McLaughlin, E. and O'Donoghue, P. (2004) Analysis of primary school children's physical activity in the playground: a complementary approach. In P.G. O'Donoghue and M.D. Hughes (eds), *Performance Analysis of Sport 6* (pp.233–40), Cardiff: CPA Press, UWIC.

McNair, D.M., Lorr, M. and Droppelman, L. (1971) *Manual: Profile of mood states*, San Diego: Educational and Industrial Testing Service Inc.

McStravick, L. and O'Donoghue, P.G. (2001) 'The effect of score-line on performance: a time-motion and sports psychology investigation in Irish League soccer', paper presented at the 3rd Exercise and Sports Science Association of Ireland Annual Conference, Carlow, April.

Marshall, C. and Rossman, G.B. (1999) *Designing Qualitative Research*, 3rd edn, London: Sage.

Martin, M. (1907) 'Le Championnay de France de rugby', Available <http://rugby-pioneers.blogs.com/rugby/2006/09/statistical_ana.html> (accessed 21 May 2009).

Martin, G.D, Murphy, M.H., O'Donoghue, P.G. and Bleakley, E.W. (1996) 'Validation of distance estimation in computerised time-motion analysis for association football', *Journal of Sports Sciences*, 15: 18.

Martin, D., Cassidy, D. and O'Donoghue, P.G. (2004) 'The effectiveness of performance analysis in elite Gaelic football', paper presented at the World Congress of Performance Analysis of Sport 6, June.

Maslovat, D. and Franks, I.M. (2008) The need for feedback. In M. Hughes and I.M. Franks (eds), *Essential of Performance Analysis: An introduction* (pp. 1–7). London: Routledge.

Massey, C., Maneval, L., Phillips, J., Vincent, J., White, G. and Zoeller, B. (2002) 'An analysis of teaching and coaching behaviours of elite strength and conditioning coaches', *Journal of Strength and Conditioning Research*, 16: 456–60.

Mayes, A., O'Donoghue, P.G., Garland, J. and Davidson, A. (2009) 'The use of performance analysis and internet video streaming during elite netball preparation', paper presented at the 3rd International Workshop of the International Society of Performance Analysis of Sport, Lincoln, April.

Mellick, M. (2005) Elite referee decision communication: developing a model of best practice. Unpublished PhD thesis, University of Wales Institute Cardiff.

Miles, M. and Huberman, A. (1994) *Qualitative Data Analysis, Xth edn*, London: Sage.

Mizohata, J., O'Donoghue, P.G. and Hughes, M. (2009) 'Work-rate of senior rugby union referees during matches', paper presented at the 3rd International Workshop of the International Society of Performance Analysis of Sport, Lincoln, April.

Montoye, H.J., Kemper, H.C.G., Saris, W.H.M. and Washburn, R.A. (1996) *Measuring Physical Activity and Energy Expenditure*, Champaign, IL: Human Kinetics.

More, K. (2008) Notational analysis of coaching. In M. Hughes and I.M. Franks (eds), *The Essentials of Performance Analysis: An introduction* (pp. 264–76). London: Routledge.

More, K. and Franks, I.M. (1996) 'Analysis and modification of verbal coaching behaviour: the usefulness of a data driven intervention strategy', *Journal of Sports Sciences*, 14: 523–43.

Morgan, L. (2006) Does the use of performance analysis based feedback enhance the performance of league netball players? Unpublished BSc Hons thesis, University of Wales Institute Cardiff.

Morrow Jr, J.R., Jackson, A.W., Disch, J.G. and Mood, D.P. (2005) *Measurement and Evaluation in Human Performance, 3rd edn*, Champaign, IL: Human Kinetics.

MSSE (2009) 'Information for authors', Available <http://journals.lww.com/acsm-msse/_layouts/1033/oaks.journals/informationforauthors.aspx> (accessed 27 July 2009)

Murray, S. and Hughes, M. (2001) Tactical performance profiling in elite level senior squash. In M. Hughes and I.M. Franks (eds) *Performance Analysis, Sports Science and Computers* (pp. 185–94). Cardiff: CPA Press, UWIC.

Murray, S., Maylor, D. and Hughes, M. (1998) A preliminary investigation into the provision of computerised analysis feedback to elite squash players. In A. Lees, I. Maynard, M. Hughes and T. Reilly (eds), *Science and Racket Sports 2* (pp. 235–40). London: E & FN Spon.

NIFS (1989) *Northern Ireland Fitness Survey: A report by the division of physical activity and health education*, Belfast: Queens University Press.

Nigg, B.M. (2007a) Introduction. In B.M. Nigg and W. Herzog (eds), *Biomechanics of the Musculo-skeletal System*, 3rd edn (pp. 1–48). Chichester, UK: Wiley.

Nigg, B.M. (2007b) General comments about modelling. In B.M. Nigg and W. Herzog (eds), *Biomechanics of the Musculo-skeletal System,* 3rd edn (pp. 513–22). Chichester, UK: Wiley.

Nigg, B.M. (2007c), Force system analysis, In B.M. Nigg and W. Herzog (eds), *Biomechanics of the Musculo-skeletal System,* 3rd edn (pp. 523–34). Chichester, UK: Wiley.

Nigg, B.M. (2007d), Mathematically determinate systems, In B.M. Nigg and W. Herzog (eds), *Biomechanics of the Musculo-skeletal System,* 3rd edn (pp. 535–608). Chichester, UK: Wiley.

Noldus (2009) <http://www.noldus.com>, (accessed 21 May 2009).

Ntoumanis, N. (2001) *A Step-by-Step Guide to SPSS for Sport and Exercise Studies*, London: Routledge.

O'Donoghue, P.G. (1998) Time-motion analysis of work-rate in elite soccer. In M. Hughes and F. Tavares (eds), *Notational Analysis of Sport 4* (pp. 65–70). Porto: University of Porto.

O'Donoghue, P.G. (2001a) 'The most important points in Grand Slam singles tennis', *Research Quarterly for Exercise and Sport*, 72: 125–31.

O'Donoghue, P.G. (2001b) Is notational analysis research? A repeated investigation of tennis strategy. In M. Hughes and I.M. Franks (eds), *Performance Analysis, Sports Science and Computers* (pp. 147–52), Cardiff: CPA Press, UWIC.

O'Donoghue, P.G. (2002) 'Performance models of ladies' and men's singles tennis at the Australian Open', *International Journal of Performance Analysis of Sport*, 2: 73–84.

O'Donoghue, P.G. (2003) 'The effect of score line on elite tennis strategy: a cluster analysis', *Journal of Sports Sciences*, 21: 284–5.

O'Donoghue, P.G. (2004) 'Sources of variability in time-motion data; measurement error and within player variability in work-rate', *International Journal of Performance Analysis of Sport*, 4(2): 42–9.

O'Donoghue, P.G. (2005a) 'Normative profiles of sports performance', *International Journal of Performance Analysis of Sport*, 5(1): 104–19.

O'Donoghue, P.G. (2005b) 'Evaluation of computer-based predictions of the Euro 2004 soccer tournament', paper presented at the 5th International Symposium of Computer Science, Hvar, Croatia, May.

O'Donoghue, P.G. (2006a) Elite tennis strategy during tie-breaks. In H. Dancs, M. Hughes and P.G. O'Donoghue (eds), *Performance Analysis of Sport 7* (pp. 654–60). Cardiff: CPA Press, UWIC.

O'Donoghue, P.G. (2006b) Performance indicators for possession and shooting in international netball. *Performance Analysis of Sport 7* (pp. 483–503). Cardiff: CPA Press, UWIC.

O'Donoghue, P.G. (2006c) 'The effectiveness of satisfying the assumptions of predictive modelling techniques: an exercise in predicting the FIFA World Cup 2006', *International Journal of Computing Science in Sport*, 5(2): 5–16.

O'Donoghue, P.G. (2007a) 'Data mining and knowledge discovery in performance analysis: an example of elite tennis strategy', paper presented at the 6th International Symposium of Computer Science in Sport, Calgary, June.

O'Donoghue, P.G. (2007b) 'Reliability issues in performance analysis', *International Journal of Performance Analysis of Sport*, 7(1): 35–48.

O'Donoghue, P.G. (2007c) 'Fault tree analysis and its application in performance analysis of sport', paper presented at the 6th International Symposium of Computer Science in Sport, Calgary, June.

O'Donoghue, P.G. (2008a) 'Principal components analysis in the selection of key performance indicators in sport', *International Journal of Performance Analysis of Sport*, 8(3): 145–55.

O'Donoghue, P.G. (2008b) Time-motion analysis. In M. Hughes and I.M. Franks (eds), *Essentials of Performance Analysis: An introduction* (pp. 180–205). London: Routledge.

O'Donoghue, P.G. (2008c) 'Performance norms and opposition effects in Grand Slam women's singles tennis', paper presented at Satellite Symposia 5 (Sports Games, Performance and Coaching) of the European College of Sports Sciences Annual Conference, Lisbon, July.

O'Donoghue, P.G. (2009a) 'Interacting Performances Theory', *International Journal of Performance Analysis of Sport*, 9: 26–46.

O'Donoghue, P.G. (2009b) 'Opposition effects in men's singles tennis at the French Open', paper presented at the 3rd International Workshop of the International Society of Performance Analysis of Sport, Lincoln, April.

O'Donoghue, P.G. (2009c) 'Predictions of the 2007 Rugby World Cup and Euro 2008', paper presented at the 3rd International Workshop of the International Society of Performance Analysis of Sport, Lincoln, April.

O'Donoghue, P.G. and Brown, E.J. (2009) 'Sequences of service points and the misperception of momentum in elite tennis', *International Journal of Performance Analysis in Sport*, 9: 113–27.

O'Donoghue, P.G. and Cassidy, D. (2002) 'The effect of specific intermittent training on the fitness of international netball players', *Journal of Sports Sciences*, 20: 56–7.

O'Donoghue, P.G. and Ingram, B. (2001) 'A notational analysis of elite tennis strategy', *Journal of Sports Sciences*, 19: 107–15.

O'Donoghue, P.G. and Liddle, S.D. (1998) A notational analysis of time factors of elite men's and ladies' singles tennis on clay and grass surfaces. In A. Lees, I. Maynard, M. Hughes and T. Reilly (eds), *Science and Racket Sports 2* (pp. 241–6), London: E & FN Spon.

O'Donoghue, P.G. and Longville, J. (2004) Reliability testing and the use of statistics in performance analysis support: a case study from an international netball tournament. In P.G. O'Donoghue and M. Hughes (eds), *Performance Analysis of Sport 6* (pp. 1–7), Cardiff: CPA Press, UWIC.

O'Donoghue, P.G. and Murphy, M.H. (1996) 'Object modelling and formal specification during real-time system development', *Journal of Network and Computer Applications*, 19: 335–52.

O'Donoghue, P.G. and Ormsby, D. (2002) 'The effectiveness of mental imagery training in enhancing free kick taking in Gaelic football', *Journal of Sports Sciences*, 20: 70.

O'Donoghue, P.G. and Scully, D. (1999) 'A case study in the application of psychological skills training for an elite netball player', paper presented at the 2nd Exercise and Sports Science Association of Ireland annual conference, Limerick, November.

O'Donoghue, P.G. and Tenga, A. (2001) 'The effect of score line on work rate in elite soccer', *Journal of Sports Sciences*, 19: 25–6.

O'Donoghue, P.G. and Williams, J.J. (2004) 'An evaluation of human and computer-based predictions of the 2003 rugby union World Cup', *International Journal of Computer Science in Sport*, 3(1), 5–22.

O'Donoghue, P.G., Robinson, J. and Murphy, M.H. (1995) An object oriented intelligent notational analysis multimedia database system, In Murphy, J. and Stone, B. (eds), *Object Oriented Information Systems* (pp. 169–72), Heidelberg: Springer-Verlag.

O'Donoghue, P.G., Robinson, J. and Murphy, M.H. (1996a) MAVIS: a multimedia match analysis system to support immediate video feedback for coaching, In M. Hughes (ed). *Notational Analysis of Sport 3* (pp. 276–85), Cardiff: CPA Press, UWIC.

O'Donoghue, P.G., Robinson J. and Murphy, M.H. (1996b) 'A database system to support immediate video feedback for coaching', paper presented at the 14th International Conference of Applied Informatics, Innsbruck, February.

O'Donoghue, P.G., Martin, G.D. and Murphy, M.H. (1996c) 'Systematic evaluation of end-user training for time and motion analysis applications', paper presented at the *2nd International Conference of Technical Informatics*, Timisoara, November.

O'Donoghue, P.G., Boyd, M., Lawlor, J. and Bleakley, E.W. (2001) 'Time-motion analysis of elite, semi-professional and amateur soccer competition', *Journal of Human Movement Studies*, 41: 1–12.

O'Donoghue, P.G., Donnelly, O., Hughes, L. and McManus, S. (2004a) 'Time-motion analysis of Gaelic games', *Journal of Sports Sciences*, 22: 255–6.

O'Donoghue, P.G., Miniss, J. and Harty, K. (2004b) 'Time-motion analysis of ladies' soccer', *Journal of Sports Sciences*, 22: 257.

O'Donoghue, P.G., Dubitzky, W., Lopes, P., Berrar, D., Lagan, K., Hassan, D., Bairner, A. and Darby, P. (2004c) 'An evaluation of quantitative and qualitative methods of predicting the 2002 FIFA World Cup', *Journal of Sports Sciences*, 22: 513–14.

O'Donoghue, P.G., Rudkin, S., Bloomfield, J., Powell, S., Cairns, G., Dunkerley, A., Davey, P., Probert, G. and Bowater, J. (2005b) 'Repeated work activity in English FA Premier League soccer', *International Journal of Performance Analysis of Sport*, 5(2): 46–57.

O'Donoghue, P.G., Mayes, A., Edwards, K.M. and Garland, J. (2008) 'Performance norms for British National Super League netball', *International Journal of Sports Science and Coaching*, 3: 501–11.

O'Donoghue, P.G., Hughes, M.G., Rudkin, S., Bloomfield, J., Cairns, G. and Powell, S. (2005a) 'Work-rate analysis using the POWER (Periods of Work Efforts and Recoveries) System', *International Journal of Performance Analysis of Sport*, 4(1): 5–21.

O'Donoghue, P.G., Dubitzky, W., Lopes, P., Berrar, D., Lagan, K., Hassan, D., Bairner, A. and Darby, P. (2004d) 'An evaluation of quantitative and qualitative methods of predicting the 2002 FIFA World Cup', *Journal of Sports Sciences*, 22: 513–14.

Olsen, E. and Larsen, O. (1997) Use of match analysis by coaches. In T. Reilly, J. Bangsbo and M. Hughes (eds), *Science and Football 3* (pp. 209–20). London: E & FN Spon.

Paisey, T. and O'Donoghue, P.G. (2008) 'Physical education teacher behaviour: a quantitative and qualitative investigation', paper presented at the World Congress of Performance Analysis of Sport 8, Magdeburg, September.

Palut, Y. and Zanone, P.G. (2005) 'A dynamical analysis of tennis: Concepts and data', *Journal of Sports Sciences*, 23: 1021–32

Patrick, J.D. and McKenna, M.J. (1988) The CABER computer system: a review of its application to the analysis of Australian rules football. In T. Reilly, A. Lees, K. Davids and W.J. Murphy (eds), *Science and Football* (pp. 267–73). London: E & FN Spon.

Patton, M.Q. (2002) *Qualitative Research and Evaluation Methods*, 3rd edn, London: Sage.

Perham, S. and O'Donoghue, P.G. (2009) 'Relative age effect in field hockey: a quantitative and qualitative investigation', paper presented at the Physical Education and Sport in Research: Aging and Physical Activity Conference, Rydzyna, September.

Ponting, R. and O'Donoghue, P.G. (2009) 'Populations, sampling, variability and reliability in performance analysis', paper presented at the 3rd International Workshop of the International Society of Performance Analysis of Sport, Lincoln, April.

Potrac, P., Jones, R. and Armour, K. (2002) '"It's all about getting respect": the coaching behaviours of an expert English soccer coach', *Sport, Education and Society*, 7: 183–202.

Preston-Dunlop, V. (1966) *Readers in Kinetography Laban. Stepping. Series A, Book 1*, London: MacDonald & Evans.

Preston-Dunlop, V. (1967a) *Readers in Kinetography Laban. Motif writing For Dance: Introducing the Symbols. Series B, Book 1*, London: MacDonald & Evans.

Preston-Dunlop, V. (1967b) *Readers in Kinetography Laban. Motif writing For Dance: More about the Symbols. Series B, Book 2*, London: MacDonald & Evans.

Preston-Dunlop, V. (1967c) *Readers in Kinetography Laban. Motif writing For Dance: Moving with a Partner. Series B, Book 3*, London: MacDonald & Evans.

Preston-Dunlop, V. (1967d) *Readers in Kinetography Laban. Motif writing For Dance: Effort Graphs. Series B, Book 4*, London: MacDonald & Evans.

ProZone® (2007) 'Effective use of data', *Official Monthly Newsletter of ProZone®*, 8, 2.

Rahnama, N., Reilly, T. and Lees, A. (2002) 'Injury risk associated with playing actions during competitive soccer', *British Journal of Sports Medicine*, 36: 354–9.

Ramsbottom, R., Brewer, J. and Williams, C. (1988) 'A progressive shuttle run test to estimate maximal oxygen uptake', *British Journal of Sports Medicine*, 22: 141–4.

Reep, C. and Benjamin, B. (1968) 'Skill and chance in association football', *Journal of the Royal Statistical Society*, Series A 131: 581–5.

Redwood-Brown, A., O'Donoghue, P.G. and Robinson, G. (2009) 'The interaction effect of positional role and score line on work-rate in FA Premier League soccer', paper presented at the 3rd International Workshop of the International Society of Performance Analysis of Sport, Lincoln, April.

Reilly, T. and Thomas, V. (1976) 'A motion analysis of work rate in different positional roles in professional football match play', *Journal of Human Movement Studies*, 2: 87–97.

Reilly, T. and Waterhouse, J. (2009) 'Chronobiology and exercise', *Mecinia Sportiva*, 13(1): 54–60.

Richers, T.A. (1995) 'Time-motion analysis of the energy systems in elite and competitive singles tennis', *Journal of Human Movement Studies*, 28: 73–86.

Robinson, G. (2006) Performance Analysis of International Netball shooting: an integrated approach. Unpublished MSc thesis, University of Wales Institute Cardiff, UK.

Robinson, P. (1992) *HOOD: Hierarchical Object Oriented Design*, Englewood Cliffs, NJ: Prentice Hall.

Robinson, J., Murphy, M.H. and O'Donoghue, P.G. (1996) 'Notational analysis of work rate within the various positional roles for elite female hockey players', *Journal of Sports Sciences*, 14: 17.

Robinson, G. and O'Donoghue, P.G. (2007) 'A weighted kappa statistic for reliability testing in performance analysis of sport', *International Journal of Performance Analysis of Sport*, 7(1): 12–19.

Robinson, G. and O'Donoghue, P.G. (2008) 'A movement classification for the investigation of agility demands and injury risk in sport', *International Journal of Performance Analysis of Sport*, 8(1): 127–44.

Rowlinson, M. and O'Donoghue, P.G. (2009) Performance profiles of soccer players in the 2006 UEFA Champions League and the 2006 FIFA World Cup tournaments. In T. Reilly and F. Korkusuz (eds), *Science and Football 6* (pp. 229–34). London: Routledge.

Rudkin, S. and O'Donoghue, P.G. (2008) 'Time motion analysis of first class cricket fielding', *Journal of Science and Medicine in Sport*, 11: 604–7.

Salmela, J.H., Marques, M., Machado, R. and Durand-Bush, N. (2006) 'Perceptions of the Brazilian football coaching staff in preparation for the World Cup', *International Journal of Sport Psychology*, 37: 139–56.

Schmolinsky, G. (1983) *Track and Field: Textbook for coaches and sports teachers,* 2nd edn, Berlin: Sportverlag.

Schön, D.A. (1983) *The Reflective Practitioner. How professionals think in action,* London: Temple Smith.

Scully, D. and O'Donoghue, P.G. (1999) 'The effect of score line on tennis strategy in grand slam men's singles', *Journal of Sports Sciences*, 17: 64–5.

Shafizadeh, M. (2008) 'Qualitative analysis of aggressive behaviors in the adolescent, youth and adult soccer World Cups, *International Journal of Performance Analysis of Sport*, 8(3): 40–48.

Shapie, M.N.M., Oliver, J., O'Donoghue, P.G. and Tong, R. (2008) 'Distribution of fight time and break time in international Silat competition', paper presented at the World Congress of Performance Analysis of Sport 8, Magdeburg, September.

Sharp, B. (1997) 'Notational analysis. It's so simple', *BASES Newsletter*, 7(8): 4.

Shaw, J. and O'Donoghue, P.G. (2004) The effect of score line on work rate in amateur soccer. In P.G. O'Donoghue and M. Hughes (eds), *Performance Analysis of Sport 6* (pp. 84–91), Cardiff: CPA Press, UWIC.

Silverman, D. (1993) *Interpreting Qualitative Data: Methods for analysing talk, text and interaction*, London: Sage.

Silverman, D. (2005) *Doing Qualitative Research: A practical handbook,* 2nd edn, London: Sage.

Smyth, J. (1989) 'Developing and sustaining critical reflection in teacher education', *Journal of Teacher Education*, 40(2): 2–9.

Spencer, M., Lawrence, S., Rechichi, C., Bishop, D., Dawson, B. and Goodman, C. (2004) 'Time-motion analysis of elite field hockey, with special reference to repeated-sprint activity', *Journal of Sports Sciences*, 22: 843–50.

Spradley, J.P. (1979) *The Ethnographic Interview*, New York: Holt, Rinehart and Winston.

Strachan, L. and Weir, P. (2006) 'Twirling through the motions: applying motor control theory to practice', *International Journal of Sports Science and Coaching*, 1: 399–404.

Strauss, A. and Corbin, J. (1998) *Basics of Qualitative Research: Techniques and procedures for developing grounded theory,* 2nd edn, London: Sage.

Sugden, J. (2002) *Scum Airways: Inside football's underground economy*, Edinburgh: Mainstream Publishing.

Tabachnick, B.G. and Fidell, L.S. (1996) *Using Multivariate Statistics,* 3rd edn, New York: Harper Collins.

Takei, Y. (2007) 'The Roche vault performed by elite gymnasts: somersaulting technique, deterministic model, and judges' scores', *Journal of Applied Biomechanics*, 23: 1–11.

Taylor, J.B., Mellalieu, S.D., James, N. and Shearer, U.A. (2008) 'The influence o match location, quality of opposition and match status on technical performance in professional association football', *Journal of Sports Sciences*, 26: 885–95.

Teddlie, C. and Tashakkori, A. (2009) *Foundations of Mixed Methods Research*, Thousand Oaks, CA: Sage.

Thomas, J.R. and Nelson, J.K. (1996) *Research Methods in Physical Activity*, 3rd edn, Champaign, IL: Human Kinetics Publishers.

Thomas, J.R. and Nelson, J.K. (2001) *Research Methods in Physical Activity*, 4th edn, Champaign, IL: Human Kinetics Publishers.

Thomas, J.R., Nelson, J.K. and Silverman, D. (2005) *Research Methods in Physical Activity*, 5th edn, Champaign, IL: Human Kinetics Publishers.

Trudel, P., Cote, J. and Bernard, D. (1996) 'Systematic observation of youth ice hockey coaches during games', *Journal of Sport Behaviour*, 19(1): 50–65.

Underwood, G. and McHeath, J. (1977) 'Video analysis in tennis coaching', *British Journal of Physical Education*, 8: 136–8.

Unierzyski, P. and Wieczorek, A. (2004) Comparison of tactical solutions and game patterns in the finals of two Grand Slam tournaments in tennis. In A. Lees, J. Kahn and I. Maynard (eds), *Science and Racket Sports 3* (pp. 169–74). London: Routledge.

US Open (2008) 'Official website of the US Open', Available <http://2008.usopen.org> (accessed 27 July 2009).

Van den Bogert, A.G. and Nigg, B.M. (2007) General comments about modelling. In B.M. Nig and W. Herzog (eds), *Biomechanics of the Musculo-skeletal System*, 3rd edn (pp. 622–44). Chichester, UK: Wiley.

Van der Mars, H. (1989) Systematic observation. In P.W. Darst, D.B. Zakrajsek and V.H. Mancini (eds) *Analyzing Physical Education and Sport Instruction* (pp. 3–18). Champaign, IL, Human Kinetics.

Vincent, W.J. (2005) *Statistics in Kinesiology*, 3rd edn. Champaign, IL: Human Kinetics.

Warburton, E.C. (2000) 'The dance on paper: the effect of notation- use on learning and development in dance', *Research in Dance Education*, 1(2): 193–213.

Wells, J., Hughes, M. and O'Donoghue, P.G. (2009) 'Intra- and inter-observer reliability of a real-time and lapsed-time analysis of canoe and kayak slalom', paper presented at the 3rd International Workshop of the International Society of Performance Analysis of Sport, Lincoln, April.

Weiner, B. (1985) 'An attributional theory of achievement motivation and emotion', *Psychological Review*, 97: 74–84.

Wilkerson, J.D. (1997) Biomechanics. In J.D. Massengale and R.A. Swanson (eds), *A History of Exercise and Sports Science* (pp. 321–66). Champaign, IL: Human Kinetics.

Williams, J.J. (2008) Rule changes in sport and the role of notation, In M. Hughes and I.M. Franks (eds), *The Essentials of Performance Analysis: An introduction* (pp. 226–42). London: Routledge.

Williams, J.J. (2009) 'An investigation into operational definitions used within performance analysis', paper presented at the 3rd International Workshop of the International Society of Performance Analysis of Sport, Lincoln, April.

Williams, R. and O'Donoghue, P. (2005) 'Lower limb injury risk in netball: a time-motion analysis investigation', *Journal of Human Movement Studies*, 49: 315–31.

Williams, L. and O'Donoghue, P.G. (2006) Defensive strategies used by international netball teams. In H. Dancs, M. Hughes and P.G. O'Donoghue (eds), *Performance Analysis of Sport 7* (pp. 498–503). Cardiff: CPA Press, UWIC.

Wimbledon (2009) 'Official website of the Wimbledon', Available <http://www. wimbledon.org> (accessed 27 July 2009).

Winkler, W. (1988) A new approach to the video analysis of tactical aspects of soccer. T. Reilly, A. Lees, K. Davids and W. Murphy (eds), *Science and Football* (pp. 368–72). London: E & FN Spon.

Withers, R.T., Maricic, Z., Wasilewski, S. and Kelly, L. (1982) 'Match analysis of Australian professional soccer players', *Journal of Human Movement Studies*, 8: 159–76.

Wright, A. and O'Donoghue, P.G. (1999) 'The influence of performance accomplishment on elite tennis strategy within matches', paper presented at the 2nd Exercise and Sports Science Association of Ireland Annual Conference, Limerick, November.

Yeadon, M.R. and King, M.A. (2008) Computer simulation modelling in sport. In C. Payton and R. Bartlett (eds), *Biomechanical Evaluation of Movement in Sport and Exercise: BASES Guidelines* (pp. 176–205). London: Routledge.

Zuber-Skerritt, O. (1996) Emancipatory action research for organisation change and management development. In O. Zuber-Skerritt (ed), *New Directions in Action Research* (pp. 83–105). London: Falmer.

INDEX